ENERGY SECTOR STANDARD OF THE PEOPLE'S REPUBLIC OF CHINA

中华人民共和国能源行业标准

Guide for Design of Energy Dissipation and Erosion Control for Water Release Structures of Hydropower Projects

水电工程泄水建筑物消能防冲设计导则

NB/T 10392-2020

Chief Development Department: China Renewable Energy Engineering Institute
Approval Department: National Energy Administration of the People's Republic of China
Implementation Date: February 1, 2021

China Water & Power Press

中国水利水电出版社

Beijing 2024

All rights reserved. No part of this publication may be reproduced, stored in a retrieval system, or transmitted in any form or by any means—electronic, mechanical, photocopying, recording or otherwise, without prior written permission of the publisher.

图书在版编目（CIP）数据

水电工程泄水建筑物消能防冲设计导则：NB/T 10392-2020 = Guide for Design of Energy Dissipation and Erosion Control for Water Release Structures of Hydropower Projects（NB/T 10392-2020）：英文 / 国家能源局发布. -- 北京：中国水利水电出版社，2024. 6. -- ISBN 978-7-5226-2706-9

Ⅰ. TV65-65

中国国家版本馆CIP数据核字第2024SD7269号

ENERGY SECTOR STANDARD
OF THE PEOPLE'S REPUBLIC OF CHINA
中华人民共和国能源行业标准

Guide for Design of Energy Dissipation and Erosion Control
for Water Release Structures of Hydropower Projects
水电工程泄水建筑物消能防冲设计导则
NB/T 10392-2020
（英文版）

Issued by National Energy Administration of the People's Republic of China
国家能源局　发布
Translation organized by China Renewable Energy Engineering Institute
水电水利规划设计总院　组织翻译
Published by China Water & Power Press
中国水利水电出版社　出版发行
　　Tel: (+ 86 10) 68545888　68545874
　　sales@mwr.gov.cn
　　Account name: China Water & Power Press
　　Address: No.1, Yuyuantan Nanlu, Haidian District, Beijing 100038, China
　　http://www.waterpub.com.cn
中国水利水电出版社微机排版中心　排版
北京中献拓方科技发展有限公司　印刷
184mm×260mm　16 开本　5.5 印张　174 千字
2024 年 6 月第 1 版　2024 年 6 月第 1 次印刷
Price（定价）：￥900.00

Introduction

This English version is one of China's energy sector standard series in English. Its translation was organized by China Renewable Energy Engineering Institute authorized by National Energy Administration of the People's Republic of China in compliance with relevant procedures and stipulations. This English version was issued by National Energy Administration of the People's Republic of China in Announcement [2023] No. 8 dated December 28, 2023.

This version was translated from the Chinese Standard NB/T 10392-2020, *Guide for Design of Energy Dissipation and Erosion Control for Water Release Structures of Hydropower Projects*, published by China Water & Power Press. The copyright is reserved by National Energy Administration of the People's Republic of China. In the event of any discrepancy in the implementation, the Chinese version shall prevail.

Many thanks go to the staff from the relevant standard development organizations and those who have provided generous assistance in the translation and review process.

For further improvement of the English version, any comments and suggestions are welcome and should be addressed to:

China Renewable Energy Engineering Institute
No. 2 Beixiaojie, Liupukang, Xicheng District, Beijing 100120, China
Website: www.creei.cn

Translating organization:

POWERCHINA Kunming Engineering Corporation Limited

Translating staff:

YANG Jijian	MIAO Jiali	JIA Haibo	PENG Fuping
YANG Zaihong	CHENG Xiaolong	GUO Xin	ZHONG Zhihui
FANG Jingyu			

Review panel members:

JIN Feng	Tsinghua University
LIU Xiaofen	POWERCHINA Zhongnan Engineering Corporation Limited
LI Shisheng	China Renewable Energy Engineering Institute
ZHANG Ming	Tsinghua University

LUO Yongqin	Kunming University of Science and Technology
YAN Wenjun	Army Academy of Armored Forces, PLA
GUO Jie	POWERCHINA Beijing Engineering Corporation Limited
LI Zhongjie	POWERCHINA Northwest Engineering Corporation Limited
WANG Shun	Wuhan University
QI Wen	POWERCHINA Beijing Engineering Corporation Limited
SUN Xushu	China Renewable Energy Engineering Institute

National Energy Administration of the People's Republic of China

翻译出版说明

本译本为国家能源局委托水电水利规划设计总院按照有关程序和规定，统一组织翻译的能源行业标准英文版系列译本之一。2023年12月28日，国家能源局以2023年第8号公告予以公布。

本译本是根据中国水利水电出版社出版的《水电工程泄水建筑物消能防冲设计导则》NB/T 10392—2020 翻译的，著作权归国家能源局所有。在使用过程中，如出现异议，以中文版为准。

本译本在翻译和审核过程中，本标准编制单位及编制组有关成员给予了积极协助。

为不断提高本译本的质量，欢迎使用者提出意见和建议，并反馈给水电水利规划设计总院。

地址：北京市西城区六铺炕北小街2号
邮编：100120
网址：www.creei.cn

本译本翻译单位：中国电建集团昆明勘测设计研究院有限公司
本译本翻译人员：杨吉健　缪嘉莉　贾海波　彭富平
　　　　　　　　杨再宏　程晓龙　郭　鑫　钟智辉
　　　　　　　　方靖宇
本译本审核人员：
　　金　峰　清华大学
　　刘小芬　中国电建集团中南勘测设计研究院有限公司
　　李仕胜　水电水利规划设计总院
　　张　明　清华大学
　　罗永钦　昆明理工大学
　　闫文军　中国人民解放军陆军装甲兵学院
　　郭　洁　中国电建集团北京勘测设计研究院有限公司
　　李仲杰　中国电建集团西北勘测设计研究院有限公司
　　王　顺　武汉大学

齐　文　中国电建集团北京勘测设计研究院有限公司
孙旭曙　水电水利规划设计总院

国家能源局

Announcement of National Energy Administration of the People's Republic of China [2020] No. 5

National Energy Administration of the People's Republic of China has approved and issued 502 energy sector standards including *Technical Code for Real-Time Ecological Flow Monitoring Systems of Hydropower Projects* (Attachment 1) and the English version of 35 energy sector standards including *Series Parameters for Horizontal Hydraulic Hoist (Cylinder)* (Attachment 2).

Attachments: 1. Directory of Sector Standards
2. Directory of English Version of Sector Standards

National Energy Administration of the People's Republic of China

October 23, 2020

Attachment 1:

Directory of Sector Standards

Serial number	Standard No.	Title	Replaced standard No.	Adopted international standard No.	Approval date	Implementation date
...						
8	NB/T 10392-2020	Guide for Design of Energy Dissipation and Erosion Control for Water Release Structures of Hydropower Projects			2020-10-23	2021-02-01
...						

Foreword

According to the requirements of Document GNKJ [2009] No. 163 issued by National Energy Administration of the People's Republic of China, "Notice on Releasing the Development and Revision Plan of the First Batch of Energy Sector Standards in 2009", and after extensive investigation and research, summarization of practical experience, and wide solicitation of opinions, the drafting group has prepared this guide.

The main technical contents of this guide include: basic requirements, type and arrangement of energy dissipation and erosion control, design for hydraulic jump energy dissipation and erosion control, design for ski-jump energy dissipation and erosion control, design for surface flow and bucket-type flow energy dissipation and erosion control, design for inside-tunnel energy dissipation and erosion control, downstream protection design, hydraulic model test and hydraulic numerical simulation, safety monitoring design, operation and maintenance requirements.

National Energy Administration of the People's Republic of China is in charge of the administration of this guide. China Renewable Energy Engineering Institute has proposed this guide and is responsible for its routine management. Energy Sector Standardization Technical Committee on Hydropower Investigation and Design is responsible for the explanation of specific technical contents. Comments and suggestions in the implementation of this guide should be addressed to:

China Renewable Energy Engineering Institute
No. 2 Beixiaojie, Liupukang, Xicheng District, Beijing 100120, China

Chief development organization:

POWERCHINA Kunming Engineering Corporation Limited

Participating development organizations:

POWERCHINA Chengdu Engineering Corporation Limited

POWERCHINA Beijing Engineering Corporation Limited

POWERCHINA Northwest Engineering Corporation Limited

POWERCHINA Zhongnan Engineering Corporation Limited

POWERCHINA Guiyang Engineering Corporation Limited

POWERCHINA Huadong Engineering Corporation Limited

Chief drafting staff:

ZHANG Zongliang	YANG Zaihong	LIU Shaochuan	YOU Xiang
KONG Caifen	HE Qiyong	ZHANG Shaochun	LUO Yongqin
ZHANG Libing	YANG Jijian	OU Hongguang	LI Yannong
HE Wei	CHEN Ruihua	YANG Hua	ZHAO Shiming
MENG Fuqiang	LIU Lijuan	LIN Jianyong	PENG Chengjia
YAO Yuancheng	XUE Wenqiang	LU Chuanyin	ZHOU Heng

Review panel members:

DANG Lincai	SU Liqun	WANG Yiming	WANG Ke
LIU Zhanping	ZHAO Hongmin	NING Huawan	ZHAO Yonggang
HUANG Tairen	QIU Huanfeng	YANG Huaide	LI Yuetao
PENG Yu	DAI Xiaobing	LIAO Renqiang	WU Wenping
HUANG Guobing	WU Weiwei	PIAO Ling	ZHAO Yi
DU Gang	LI Shisheng		

Contents

1	**General Provisions**	1
2	**Terms**	2
3	**Basic Requirements**	5
4	**Type and Arrangement of Energy Dissipation and Erosion Control**	**8**
4.1	General Requirements	8
4.2	Selection of Energy Dissipation Types	8
4.3	Downstream Protection	10
5	**Design for Hydraulic Jump Energy Dissipation and Erosion Control**	**11**
5.1	General Requirements	11
5.2	Configuration Design	11
5.3	Hydraulic Design	13
5.4	Structural Design	14
6	**Design for Ski-Jump Energy Dissipation and Erosion Control**	**16**
6.1	General Requirements	16
6.2	Configuration Design	17
6.3	Hydraulic Design	20
6.4	Structural Design	21
7	**Design for Surface Flow and Bucket-Type Flow Energy Dissipation and Erosion Control**	**23**
7.1	General Requirements	23
7.2	Configuration and Hydraulic Design	23
7.3	Structural Design	26
8	**Design for Inside-Tunnel Energy Dissipation and Erosion Control**	**27**
8.1	General Requirements	27
8.2	Configuration Design	27
8.3	Hydraulic Design	29
9	**Downstream Protection Design**	**32**
9.1	General Requirements	32
9.2	Protection Type and Structural Design	34
9.3	Protection of Flood Discharge Atomization Zone	36
10	**Hydraulic Model Test and Hydraulic Numerical Simulation**	**37**

10.1	General Requirements		37
10.2	Hydraulic Model Test		37
10.3	Hydraulic Numerical Simulation		38
11	**Safety Monitoring Design**		**39**
11.1	General Requirements		39
11.2	Structural Safety Monitoring		39
11.3	Hydraulic Prototype Observation		40
12	**Operation and Maintenance Requirements**		**42**
Appendix A	**Hydraulic Calculation for Hydraulic Jump Energy Dissipation**		**43**
Appendix B	**Configuration Parameters for End-Flared Pier**		**49**
Appendix C	**Dissipator Floor Stability Calculation**		**52**
Appendix D	**Hydraulic Calculation for Ski-Jump Energy Dissipation**		**56**
Appendix E	**Hydraulic Calculation for Surface Flow Energy Dissipation**		**63**
Appendix F	**Hydraulic Calculation for Inside-Tunnel Energy Dissipation**		**66**
Appendix G	**Scour Calculation of Downstream River Channel**		**70**
Appendix H	**Empirical Allowable Non-scouring Velocity for Downstream Channel**		**73**
Explanation of Wording in This Guide			**75**
List of Quoted Standards			**76**

1 General Provisions

1.0.1 This guide is formulated with a view to standardizing the layout, configuration design, hydraulic design, structural design, and operation and maintenance of energy dissipation and erosion control works for water release structures of hydropower projects to achieve the objectives of operation safety, technological advancement, and economic rationality.

1.0.2 This guide is applicable to the design of energy dissipation and erosion control for water release structures of hydropower projects.

1.0.3 The design of energy dissipation and erosion control for water release structures of small hydropower projects may be simplified on the basis of this guide.

1.0.4 The design of energy dissipation and erosion control for water release structures of hydropower projects in plain regions may comply with the current sector standard SL 265, *Design Specifications for Sluices*.

1.0.5 In addition to this guide, the design of energy dissipation and erosion control for water release structures of hydropower projects shall comply with other current relevant standards of China.

2 Terms

2.0.1 water release structure

hydraulic structure used for discharging the flood or waterlog that exceeds the regulation or storage capacity of a reservoir, river channel, canal, or waterlogging area, or for discharging the stored water in a reservoir or canal for safety protection or inspection and maintenance

2.0.2 hydraulic structure of energy dissipation and erosion control

hydraulic structure arranged along a water release structure or at its outlet to make the high-velocity discharge well connected with the downstream channel and form a local hydraulic energy dissipation flow pattern such as trajectory nappe, hydraulic jump, whirling, drastic contraction and diffusion, and spiral flow, so as to prevent scouring of the downstream riverbed and bank slopes

2.0.3 stilling basin

energy dissipation facility arranged downstream of a dam, sluice, or water release structure, which is composed of sidewalls, floor, end sill, apron, etc., according to the requirements of hydraulic jump energy dissipation

2.0.4 plunge pool

structure arranged downstream of a dam or spillway to provide a water area big and deep enough to meet the requirements of ski-jump and drop energy dissipation

2.0.5 combined energy dissipation

energy dissipation and erosion control arrangement where multiple energy dissipators are used for a single water release structure, or two or more water release structures work together to make the discharges collide, impact and shear to intensify the water turbulence and energy dissipation, hence improving the energy dissipation efficiency, reducing the energy dissipator scale, and reducing the downstream scouring

2.0.6 outlet energy dissipation

energy dissipation and erosion control arrangement at the outlet or end of a water release structure to dissipate the discharge energy by hydraulic jump, ski-jump, surface flow, etc.

2.0.7 inside-tunnel energy dissipation

energy dissipation and erosion control arrangement in a spillway tunnel to dissipate discharge energy using thick orifice, orifice plate, swirl, etc.

2.0.8 hydraulic jump energy dissipation

energy dissipation method in which hydraulic jump is used to dissipate the residual energy of rapid flow discharged from the bottom of a water release structure by transforming rapid flow into slow flow to connect the downstream flow

2.0.9 ski-jump energy dissipation

energy dissipation method in which a flip bucket is set at the outlet of a water release structure to deflect rapid flow into air, forming aerated jet which falls into the downstream water cushion

2.0.10 surface flow energy dissipation

energy dissipation method in which a drop sill or bucket is set at the end of a water release structure to deflect the main stream of the discharged rapid flow to the water surface, dissipating the residual energy by surface dispersion of the main stream, bottom swirling, and surface swirling

2.0.11 bucket-type flow energy dissipation

energy dissipation method in which a bucket is set at the end of a water release structure to deflect the discharge to the water surface, dissipating energy by swirling inside the bucket, bottom swirling, and surface swirling, also called roller bucket energy dissipation

2.0.12 end-flared pier

overflow dam pier whose rear part is widened as fishtail

2.0.13 auxiliary energy dissipator

auxiliary structure which collaborates with the main dissipator by partially changing the discharge flow pattern or flow morphology and intensifying turbulent shear to improve the energy dissipation efficiency

2.0.14 drop energy dissipation

energy dissipation method in which the flow from the outlet of water release structure falls directly into the water cushion

2.0.15 shaft swirl energy dissipation

energy dissipation method in which a swirl chamber is set in the shaft to spin and aerate the flow against the shaft wall down to the deep water cushion at the shaft bottom

2.0.16 horizontal vortex energy dissipation

energy dissipation method in which a swirl chamber is set at the bottom of a pressurized shaft to spin the flow against the wall of a horizontal tunnel before entering a plunge pool

2.0.17 orifice plate energy dissipation

energy dissipation method in which orifice plates are arranged at specified intervals in a pressurized spillway tunnel to reduce the sectional area of the tunnel locally, resulting in sudden contraction and dispersion of the water flow

2.0.18 thick orifice energy dissipation

energy dissipation method in which one or multiple thick orifices are arranged in a pressurized spillway tunnel to intensify flow turbulence by sudden contraction and expansion of sections

3 Basic Requirements

3.0.1 The design of energy dissipation and erosion control for water release structures of hydropower projects shall include the layout, configuration design, hydraulic design, structural design, and safety monitoring design of energy dissipation and erosion control structures, downstream river channel protection works, and bank slope protection works.

3.0.2 The following data shall be collected and analyzed for the design of energy dissipation and erosion control for water release structures of hydropower projects:

1. Natural conditions such as meteorology, hydrology, sediment, topography, and geology.

2. Requirements of reservoir operation, social and ecological environment, flood control, navigation, etc.

3.0.3 Classification and flood standards of hydraulic structures of energy dissipation and erosion control shall comply with the current sector standard DL 5180, *Classification & Design Safety Standard of Hydropower Projects*.

3.0.4 The hydraulic design of energy dissipation and erosion control structures shall meet the following requirements:

1. Stable energy dissipation flow pattern and good energy dissipation performance shall be achieved under the design flood, and hydrodynamics characteristics shall be desirable under frequent floods.

2. The outflow shall be coordinated with the adjacent structures and the requirements of downstream protection and navigation, and shall not affect the normal operation of other hydropower complex structures and important downstream facilities.

3. The maintenance conditions during operation should be considered.

4. The preliminary hydraulic design may be conducted based on the results of hydraulic numerical analysis when the hydraulic conditions are complex or a novel energy dissipator is employed.

3.0.5 The structural design of energy dissipation and erosion control structures shall meet the following requirements:

1. The structural configuration, material properties, foundation anchoring, and foundation treatment shall be designed according to the hydraulic

characteristics and the requirements for stability against floating and sliding.

2 For concrete structures, structural joints should be widely spaced. Reliable waterstops shall be arranged and enclosed at the surface of structural joints, and keys should be arranged in the structural joints of the floor. Clear technical requirements for the treatment of construction joints shall be proposed.

3 For concrete structures, temperature control and crack prevention design should be performed.

4 Seepage control and drainage measures shall be taken according to the engineering geological and hydrogeological conditions and the operation and maintenance requirements. An enclosed pumping system should be set in the stilling basin or plunge pool.

5 The characteristic value of anchorage force of foundation anchor bars shall be calculated in accordance with the effective weight requirements of anchored foundation specified in the current sector standard DL/T 5166, *Design Specification for River-Bank Spillway*, and the sectional area and length of anchor bars shall be checked in accordance with the current national standard GB 50086, *Technical Code for Engineering of Ground Anchoring and Shotcrete Support*.

6 Foundation treatment measures shall be proposed for bedrock.

3.0.6 The design of resistance against abrasion and cavitation for hydraulic structures of energy dissipation and erosion control shall consider the sediment characteristics, hydraulic characteristics, structure, raw materials of concrete, temperature and crack control requirements, construction technology, etc., and shall meet the following requirements:

1 The structural configuration shall be determined reasonably, and aeration and erosion reduction facilities shall be set.

2 The surface unevenness of flow passages shall comply with the current sector standard DL/T 5166, *Design Specification for River-Bank Spillway*.

3 The strength class of abrasion- and cavitation-resistant concrete shall be determined by the flow velocity, which shall not be inferior to C30 but should not be superior to C50.

4 The concrete of a dissipator floor should be placed continuously. Where abrasion-resistant concrete is required for the surface layer of the floor,

the concrete shall be placed continuously together with the concrete below the surface layer.

5 Novel dissipators should be subjected to depressurization model test.

4 Type and Arrangement of Energy Dissipation and Erosion Control

4.1 General Requirements

4.1.1 The layout of hydraulic structure of energy dissipation and erosion control shall be determined through techno-economic comparison according to the project purpose, reservoir operation requirements and environmental conditions, considering topographical and geological conditions, project layout, outflow connection, downstream erosion resistance, bank slope stability, construction conditions, operation and maintenance, project investment, etc.

4.1.2 For a large- or medium-sized hydropower project with complex hydraulic conditions, or a water release structure adopting combined energy dissipation or novel dissipators, hydraulic model tests shall be conducted for the arrangement of energy dissipation and erosion control.

4.2 Selection of Energy Dissipation Types

4.2.1 Depending on the type of water release structure and the location of dissipator, the energy dissipation may adopt outlet energy dissipation or inside-tunnel energy dissipation.

4.2.2 The outlet energy dissipation may use hydraulic jump, ski-jump, drop flow, surface flow, bucket-type flow and other energy dissipation methods, and should be selected according to the following requirements:

1. Hydraulic jump energy dissipation should be adopted when there are navigation requirements, unfavorable geological bodies or atomization-sensitive factors downstream.

2. Ski-jump energy dissipation should be adopted when the head is high or medium and the scour hole would not threaten the safety of structures and bank slopes.

3. Surface flow or bucket-type flow may be adopted when the tailwater is deeper with a minor water level fluctuation.

4.2.3 For hydraulic jump energy dissipation, the following requirements shall be met:

1. The use of auxiliary dissipators such as end-flared pier, stepped spillway, and drop sill should be studied when the gravity dam is higher than 100 m.

2. Hydraulic jump energy dissipation should be demonstrated when the

dam is higher than 150 m or the velocity of flow entering stilling basin is over 40 m/s.

4.2.4 For ski-jump energy dissipation, the following requirements shall be met:

1. The arrangement and shape of a flip bucket shall be determined according to the topography, geology, and suitability of the downstream valley to disperse the jet and prevent scouring the structure foundation and bank slope.

2. The structural measures such as plunge pool, end dam, bank protection, bottom protection shall be studied when the scour hole would threaten the safety of structures and bank slopes.

4.2.5 Inside-tunnel energy dissipation may adopt swirling flow, orifice plate, thick orifice, etc., and the selection shall meet the following requirements:

1. Free flow connection should be adopted downstream the tunnel section of swirling flow energy dissipation.

2. Orifice plates and thick orifices shall be set in the pressure flow section of the tunnel.

4.2.6 For the project with a large discharge or high head or in a narrow valley, when the single energy dissipation is difficult to arrange or is not effective, the combined energy dissipation may be adopted, and the following requirements shall be met:

1. For one water release structure, the auxiliary dissipator may be added to improve the energy dissipation efficiency, or two or more dissipators may be used for combined energy dissipation.

2. For different water release structures, collision, impact, and shear of jets may be adopted to improve the energy dissipation efficiency.

3. Distributed arrangement and zoned energy dissipation may be adopted.

4. When an overflow dam adopts hydraulic jump or bucket-type flow energy dissipation, end-flared piers may be adopted, or steps may be set on the spillway surface.

5. When ski-jump or drop energy dissipation is adopted, the overflow spillway in an arch dam may work together with the high-level or bottom outlet to form a joint energy dissipation by horizontal collision or vertical impact, while the impact of flood discharge atomization on other structures shall be fully considered. Side contraction without

collision may be adopted when there are atomization restraints downstream.

4.3 Downstream Protection

4.3.1 The type and range of protection for downstream riverbed and bank slopes at the energy dissipator outlet shall be determined according to the topographical and geological conditions, energy dissipation methods, river hydraulic conditions, or the protection requirements of affected objects downstream.

4.3.2 When ski-jump energy dissipation is used, the impacts of atomization on the surroundings, bank slope stability, traffic conditions, and operation of other structures shall be analyzed, and protection measures shall be taken accordingly.

5 Design for Hydraulic Jump Energy Dissipation and Erosion Control

5.1 General Requirements

5.1.1 The design for hydraulic jump energy dissipation and erosion control shall include the following:

1. The configuration design of stilling basin and auxiliary energy dissipators.

2. The hydraulic design for hydraulic jump form, water surface, flow velocity, hydrodynamic pressure of stilling basin.

3. The structural design of energy dissipator, including the concrete material properties and temperature control, structural joints and waterstops, floor floating stability, sidewall overall stability, foundation anchoring, structural reinforcement, foundation treatment, seepage control and drainage, etc.

5.1.2 The type of auxiliary energy dissipator, such as end-flared pier, stepped spillway, drop sill, and small deflector, shall be determined considering the factors such as design head, unit discharge, downstream water depth, anti-abrasion, and anti-cavitation.

5.2 Configuration Design

5.2.1 The configuration design of stilling basin should meet the following requirements:

1. The plan of stilling basin should be symmetrical, straight and constant in width. When the unit discharge is large, the scour resistance of bedrock is poor, or the downstream water is shallow, the gradual expansion design may be adopted.

2. The longitudinal section of a rectangular stilling basin may be determined by the hydraulic calculation for hydraulic jump energy dissipation, which should be in accordance with Appendix A of this guide. When the downstream water depth is insufficient, measures such as excavation and end sill may be taken to increase the water depth in the basin.

3. The cross section of stilling basin should be coordinated with the requirements of bank slope excavation and sidewall structural configuration. The rectangular or trapezoidal section should be adopted.

5.2.2 The configuration design of end-flared piers shall meet the following requirements:

1. The configuration parameters of end-flared piers may be preliminarily selected in accordance with Appendix B of this guide and engineering experience, and shall be finalized by hydraulic model tests.

2. The configuration of end-flared piers shall be coordinated with the layout of the gate chamber of the overflow dam, and shall not affect the discharge capacity and the arrangement of the hinged support of radial gate.

3. The end-flared piers of side orifice should be asymmetric, and the contracted nappe shall not go beyond the downstream sidewall.

5.2.3 When stepped spillway works jointly with end-flared piers, the following requirements shall be met:

1. The nappe at the bottom of end-flared piers shall form an air cavity. A small deflector may be set before the first step of stepped spillway, and the height of first several steps should be 1.5 m to 2.0 m.

2. The connection of stepped spillway and stilling basin floor should be arc or small jet angle.

5.2.4 When the velocity of flow entering stilling basin is over 30 m/s, a drop sill should be set at the entrance of stilling basin to decrease the velocity of flow approaching the basin floor, and the configuration design of drop sill should meet the following requirements:

1. The angle and height of drop sill may be preliminarily determined by the upstream and downstream hydraulic conditions and similar engineering experience, and should be finalized by hydraulic model tests.

2. The sill top at the basin entrance should be horizontal or of small depression angle, and the upstream of sill should be connected to the abutments or chute through a reverse curve section, while a certain length of straight line section shall be set from the end of arc to the sill top.

3. The height of drop sill shall be determined according to factors such as flow velocity at basin entrance, unit discharge, downstream water depth, etc., considering the angle at basin entrance and end sill height, the control requirements of hydraulic characteristics such as flow pattern, velocity of flow approaching the basin floor, and fluctuation

pressure.

4   Both sidewalls of the drop sill should be suddenly enlarged at the same section without causing cavitation damage to the downstream sidewalls.

5.3 Hydraulic Design

5.3.1 The hydraulic design of stilling basin shall meet the following requirements:

1   The configuration of stilling basin shall ensure the stable submerged hydraulic jump under different design discharges. The submergence ratio of hydraulic jump should be 1.05 to 1.10.

2   The elevation of the sidewall top shall be determined by considering the post-jump depth under design flood and an appropriate margin.

3   Auxiliary energy dissipators such as baffle piers should not be set to avoid cavitation damage when the velocity of flow approaching the basin floor is over 15 m/s.

4   When the stilling basin is connected to the steep chute, the conjugate depth may be calculated according to conjugate depth ratio of Article A.0.5 of this guide to identify the hydraulic jump type.

5   The secondary hydraulic jump should not occur downstream of the end sill of the stilling basin. The discharge velocity, water surface fluctuation, and scouring and silting form shall meet the requirements for power generation, navigation, and bank slope stability.

5.3.2 The hydraulic design of a hydraulic jump stilling basin with end-flared piers shall meet the following requirements:

1   The length may be preliminarily taken as 2/3 of the length of regular flat-floor stilling basin with two-dimensional hydraulic jump.

2   For different design flows, the stable three-dimensional hydraulic jump shall be formed in the basin.

3   The adequate dynamic water cushion depth shall be provided at the location of nappe impingement.

4   The contracted nappe after pier shall not impinge on the spillway surface or stilling basin sidewalls.

5   For a stepped spillway, sufficient aeration shall be ensured.

6   The flow pattern of spillway surface and stilling basin, energy

dissipation effectiveness, aeration and scour reduction performance, and hydrodynamic pressure characteristics should be verified by hydraulic model tests.

5.3.3 The hydraulic design for hydraulic jump stilling basin of drop sill shall meet the following requirements:

1. The floor elevation and the longitudinal section configuration may be determined according to the hydraulic calculation formulae or numerical simulation of regular flat-floor stilling basin.

2. For different design flows, the stable submerged hydraulic jump shall be formed in the basin.

3. The jet length shall be adequate in the basin to ensure sufficient diffusion and shear.

4. The flow pattern, flow velocity, hydrodynamic pressure and energy dissipation performance of drop sill stilling basin should be verified by hydraulic model tests. The depressurization model test should be conducted for the drop sill and its downstream basin floor, and the cavitation characteristics of the sidewalls.

5.4 Structural Design

5.4.1 The design of resistance against abrasion and cavitation for stilling basin shall meet the requirements of Article 3.0.6 of this guide.

5.4.2 The requirements for concrete temperature control of stilling basin shall be proposed according to the climatic characteristics, construction conditions, structure characteristics, materials properties, etc.

5.4.3 The structural stability check of stilling basin shall meet the following requirements:

1. The basin floor shall be subjected to stability check. The stability check of dissipator floor shall meet the requirements of Appendix C of this guide.

2. The stability, foundation stresses, and section stresses of sidewalls shall be checked.

3. The flow induced vibration model test should be conducted for the stilling basin with complex hydraulic characteristics or a new configuration.

5.4.4 The concrete structural jointing of stilling basin floor or sidewalls in the hydraulic jump zone should meet the following requirements:

1   The spacing of structural joints parallel to flow direction may be 15 m to 20 m.

2   The structural joints normal to flow direction should be minimized, and construction joints may be set according to the requirements of construction and temperature control.

3   Adjacent structural joints may be staggered.

4   Keys should be provided in the structural joints, and no elastic material should be filled in the joints.

5.4.5   The waterstops of stilling basin structural joints shall meet the following requirements:

1   The copper waterstops should not be less than 2 lines in the structural joints.

2   One line of copper waterstops should be provided for construction joints.

3   The construction technical requirements shall be proposed for the waterstop material properties, connection quality, and vibration of concrete in the vicinity of waterstop.

5.4.6   The design of anchor bars for stilling basin foundation shall meet the requirements of Article 3.0.5 of this guide. The exposed end of foundation anchor bars shall be anchored to the floor concrete and should be connected with the upper surface reinforcing mesh of the floor.

5.4.7   The stilling basin floor and sidewalls should be rested on weakly weathered bedrock. In structural design, foundation treatment measures such as consolidation grouting, excavate-and-replace shall be taken according to the bedrock intactness and the development of discontinuities.

5.4.8   A drainage system should be provided for the stilling basin floor. When the enclosed drainage system is adopted, grout curtains should be set.

6 Design for Ski-Jump Energy Dissipation and Erosion Control

6.1 General Requirements

6.1.1 The design for ski-jump energy dissipation and erosion control shall include:

1. The configuration design of flip bucket, plunge pool, and end dam.

2. The hydraulic calculation of various typical conditions, including the length and impact area of jets, maximum depth of scour hole, hydrodynamic pressure, etc.

3. The overall stability and floating stability of ski-jump energy dissipation structure, and the design of integral structure and structural members, foundation treatment, anchorage, drainage, structural jointing, and waterstop.

4. The design of resistance against abrasion and cavitation related to high velocity flow.

5. Other relevant design.

6.1.2 In addition to Article 4.2.4 of this guide, the design for ski-jump energy dissipation and erosion control shall also meet the following requirements:

1. The underside of nappe shall be well aerated to ensure its stability.

2. The flip bucket should be arranged on the site with foundation with favorable geological conditions and stable bank slopes.

3. The lining type of plunge pool shall be determined according to the hydraulic conditions of inflow, water cushion depth, and topographical and geological conditions of riverbed, including excavation without lining, slope protection without bottom protection, and concrete lining. When there is a sufficient water cushion in the downstream to withstand the impact of discharging flow, the lining may be simplified after verification by hydraulic model tests.

4. The plunge pool fully lined with concrete shall be arranged on stable rock foundation, and the adverse effects of rock mass structural features and geological structure on the stability of plunge pool and slopes shall be considered.

5. If unstable rock blocks exist on the bank slopes of plunge pool, they shall be removed or reinforced in time, and hard solid materials such as

rubbles and debris shall be prevented from entering or remaining in the plunge pool.

6 The plunge pool should have maintenance conditions.

6.1.3 Treatment measures shall be taken for ski-jump energy dissipation in the following cases:

1 Gently dipped weak discontinuity or fault fractured zone extending downstream in the foundation might be sheared and ruptured by the scour hole and thus threaten the safety of dam or flip bucket.

2 Bank slopes might collapse due to flood discharge, endangering the stability of abutments and blocking the outlet channel or downstream river channel.

3 Downstream surge and backflow endanger the safety and normal operation of the dam and other structures.

6.2 Configuration Design

6.2.1 The plan of flip bucket may be constant in width, expanded, or contracted. The configuration of flip bucket may be solid, slotted, or of other special shapes, and it shall be determined after comparison. The elevation of flip bucket shall ensure a free jet.

6.2.2 The configuration design of solid flip buckets shall meet the following requirements:

1 The trajectory angle shall be comprehensively determined by the conditions of flow returning to the main channel, scouring situation of scour hole and river banks, and flow pattern.

2 The bucket radius shall be comprehensively determined considering the chute slope, flow velocity and unit discharge in the bucket invert, and trajectory angle. The bucket radius may be 6 to 12 times the flow depth at the lowest point of bucket under check flood. When the chute slope is steep, or the flow velocity or unit discharge in the bucket invert is relatively large, the bucket radius should adopt the larger value.

3 According to the requirements of flow returning to the main channel, a horizontal section or flat adverse slope may be set downstream the bucket lip or the end of bucket invert for connection.

6.2.3 The configuration design of slotted buckets shall meet the following requirements:

1 The trajectory angle difference between the tooth and the slot should be

5° to 10°, the ratio of widths should be greater than 1, and the elevation difference may be 1.5 m. The configuration shall be finalized by hydraulic model tests.

    2   The cavitation damage of bucket shall be analyzed. Aerated holes should be arranged on the sides of teeth. The corners of top surface should be rounded.

6.2.4   The configuration design of slit-type flip buckets shall meet the following requirements:

    1   The outlet cross section may be symmetrically rectangular, trapezoidal, Y-shaped, V-shaped, etc., or asymmetrical, which shall be determined after comparison.

    2   The trajectory angle of bucket floor should be 0° or small, positive or negative.

    3   The contraction ratio may be 0.125 to 0.500. Under the conditions of large discharge and low Froude number, the length to width ratio of contracted section should be 0.750 to 1.500, and the corresponding contraction ratio shall take the higher value. In the case of high Froude number, the length to width ratio of contracted section should be 1.500 to 3.000, and the corresponding contraction ratio shall take the lower value.

    4   The hydrodynamic loads on outlet sidewalls shall be analyzed.

6.2.5   When the downstream valley is narrow, or the angle between the spillway axis and the river centerline is large, the special-shaped flip bucket may be adopted to control the nappe direction, plunge point, and scour hole location. The special-shaped flip bucket may be a distorted type flip bucket, curved flip bucket, chamfer type flip bucket, tongue-shaped flip bucket, swallowtail-shaped flip bucket, and flip bucket of asymmetric contraction or expansion of sidewalls. The configuration of special-shaped flip bucket shall be verified by hydraulic model tests, and the depressurization model test should be conducted.

6.2.6   The configuration design of plunge pool shall meet the following requirements:

    1   The depth of plunge pool shall meet the following requirements:

        1)   The elevation of plunge pool floor shall be determined according to the hydraulic and geological conditions, and should not be lower than the foundation surface of the dam.

2) The inflow shall form submerged hydraulic jump.

3) The hydrodynamic pressure and fluctuation pressure shall be controlled within the bearing capacity of the floor.

2 The length of plunge pool shall not be less than the sum of jet trajectory length and submerged hydraulic jump length, and should meet the following requirements:

1) The outflow from the plunge pool should approach to normal slow flow.

2) The hydrodynamic pressure on the upstream face of the end dam should be close to hydrostatic pressure.

3 The width of plunge pool shall be comprehensively determined considering the topographical and geological conditions, width of nappe entering the plunge pool, water flow pattern in the pool, and hydrodynamic pressure distribution.

4 The design of plunge pool sidewalls shall meet the following requirements:

1) The sidewall top shall be higher than the design flood level of the plunge pool, and a certain freeboard shall be considered.

2) The access for plunge pool maintenance should be considered for determining the top width of sidewalls.

3) The sidewalls of plunge pool shall be connected to the slope protection measures in the atomization area.

5 The cross section of plunge pool may be a trapezoid, compound trapezoid, or inverted arch. The flat floor of trapezoid section should be connected to the sidewalls via an arc.

6.2.7 The configuration design of end dams shall meet the following requirements:

1 The sectional size of end dam shall be designed according to the operation and maintenance conditions of plunge pool.

2 The height of end dam shall be determined according to the water cushion depth of plunge pool and maintenance requirements. The scour downstream of the end dam shall not affect the stability of foundation and bank slopes.

3 The crest width of end dam shall facilitate the maintenance of plunge

pool, and the crest should be connected to the downstream dam slope via an arc.

6.3 Hydraulic Design

6.3.1 The hydraulic calculation of ski-jump energy dissipation shall be conducted for each typical flow and starting ski-jump flow in accordance with Appendix D of this guide, and shall meet the following requirements:

1. The determination of safety trajectory length and jet width at plunge point shall not affect the stability of bedrock at dam toe, and the safety of flip bucket foundation, bank slopes and adjacent structures.

2. The upstream slope of scour hole shall be estimated according to the geological conditions, or may be taken as 1 : 3 to 1 : 6.

3. The scour to bucket foundation shall be analyzed under low discharge.

4. The starting ski-jump flow should be less than the minimum operation flow of the water release structure.

6.3.2 The flow-cavitation index of each part of the bucket should be greater than the initial cavitation index. The initial cavitation index may be determined by depressurization model test or engineering analogy. The prediction of cavitation damage may be performed in accordance with the current sector standard DL/T 5166, *Design Specification for River-bank Spillway*.

6.3.3 The height of flip bucket sidewalls shall be determined according to the maximum flow surface fluctuation before starting ski-jump and the calculated flow surface profile after aeration with a freeboard of 0.5 m to 1.5 m. For the parts with complex hydraulic conditions, the freeboard should be appropriately increased under possible unfavorable operation conditions. The water depth in fluctuation and after aeration shall be calculated in accordance with the current sector standard DL/T 5166, *Design Specification for River-bank Spillway*.

6.3.4 When the equal-width solid flip bucket is adopted, the trajectory length, maximum depth and extent of water cushion at the scour hole may be estimated according to Section D.1 of this guide.

6.3.5 When the slit-type flip bucket is adopted, the trajectory length of inner and outer edge of nappe, the maximum depth and extent of water cushion may be estimated by the method specified in Section D.2 of this guide. The intersection point of impact waves may be estimated in accordance with the current sector standard DL/T 5166, *Design Specification for River-bank Spillway*, and shall meet the following requirements:

1. Hydraulic jump shall be avoided in the contracted section.

2 The flip bucket width may be contracted once or several times.

3 The intersection point of impact waves caused by sidewall contraction should be close to the outlet of flip bucket.

6.3.6 The hydraulic design of plunge pool shall determine the length, width, location, and hydrodynamic pressure of different flows entering the plunge pool. The hydraulic design should be verified by hydraulic model tests.

6.4 Structural Design

6.4.1 The structural design of flip bucket shall include overall stability analysis, foundation stress calculation, structural member design, detailing, etc.

6.4.2 The flip bucket should rest on rock foundation. The foundation treatment design shall be determined according to the geological conditions and foundation stress analysis.

6.4.3 The temperature effects shall be considered for the concrete structure of flip bucket, and structural and construction measures shall be taken according to the local climate, structural characteristics, and foundation constraint.

6.4.4 For a flip bucket, the structural joints normal to flow direction should not be set. The construction joints shall be spaced according to the climate characteristics, foundation constraint, concrete construction and temperature control requirements.

6.4.5 The hydrodynamic pressure acting on the floor and sidewalls of flip bucket during flood discharge may be calculated in accordance with the current national standard GB/T 51394, *Standard for Load on Hydraulic Structures*. For important or complex bucket structure, the hydrodynamic pressure should be determined through hydraulic model tests and comprehensive analysis.

6.4.6 The structural design of flip bucket shall comply with the current sector standard DL/T 5057, *Design Specification for Hydraulic Concrete Structures*. The finite element method should be applied to analyze the complex bucket structure.

6.4.7 The flip bucket structure shall meet the strength and rigidity requirements, and the vibration analysis should be performed and vibration control measures should be taken for complex bucket structures.

6.4.8 The thickness of sidewalls and floor lining concrete of plunge pool shall be determined according to the geological and hydraulic conditions. The lining block size shall be determined according to the floating stability and temperature control requirements.

6.4.9 The structural design of the floor and sidewalls of plunge pool shall meet the requirements of Article 3.0.5 of this guide. The stability check of plunge pool floor shall meet the requirements of Section C.3 of this guide.

6.4.10 The enclosed pumping and drainage facility should be arranged for the plunge pool lining subjected to a higher uplift pressure, and shall meet the following requirements:

   1 The enclosed pumping and drainage system may consist of impervious curtain, drain hole, blind drainage ditch, drainage gallery, sump, pump house, and other facilities.

   2 Waterstops shall be set in the structural joint of plunge pool lining, and a drainage gallery or blind drainage ditch shall be arranged at the bottom of the structural joint.

   3 An independent pumping system for maintenance should be set.

6.4.11 When abrasion-resistant concrete is applied at the surface of plunge pool floor and sidewalls, flip bucket, and drop sill, the requirements of Article 3.0.6 of this guide shall be met.

6.4.12 If the end dam is a gravity dam, its layout and structural design shall comply with the current sector standard NB/T 35026, *Design Code for Concrete Gravity Dams*.

7 Design for Surface Flow and Bucket-Type Flow Energy Dissipation and Erosion Control

7.1 General Requirements

7.1.1 The design for surface flow and bucket-type flow energy dissipation and erosion control shall include:

1 The configuration design of flip bucket and auxiliary energy dissipator.

2 The hydraulic calculation under various flow conditions.

3 The structural design of energy dissipator.

7.1.2 When surface flow or bucket-type flow energy dissipation is combined with other energy dissipators, the following requirements shall be met:

1 The combined energy dissipation scheme shall be determined through comprehensive techno-economic comparison, considering the topographical and geological conditions, project layout, operating conditions, working head, unit discharge, downstream water depth, etc.

2 Combined energy dissipators may be used for high and medium dams with a large flood discharge, relatively narrow riverbed, poor downstream geological conditions, or poor economic rationality of single energy dissipation type. The combined energy dissipators shall be verified by hydraulic model tests.

7.1.3 When the flow pattern of surface flow or bucket-type flow energy dissipation is complex, the following engineering protection measures should be taken:

1 Set a key wall at the bottom of the bucket.

2 Set an apron downstream the bucket.

3 Set baffle piers in the energy dissipation area.

4 Set guide walls on both sides of bucket downstream.

5 Provide bank revetment downstream.

7.2 Configuration and Hydraulic Design

7.2.1 The configuration design for surface flow or bucket-type flow energy dissipation shall comprehensively consider the project operation requirements, topographical and geological conditions, working head, downstream water depth, water level change, etc.

7.2.2 The hydraulic calculation of surface flow energy dissipation should meet the requirements of Appendix E of this guide. Configuration parameters of surface flow energy dissipation, such as bucket lip height a, trajectory angle θ, reverse arc radius R, and drop sill length L (Figure 7.2.2), shall meet the following requirements:

1. The bucket lip height a should be determined according to the required flow pattern and should be greater than the minimum value a_{min}.

2. The bucket trajectory angle θ should be 0° to 25°.

3. The reverse arc radius R should not be less than 2.5 times the contracted water depth h_1 on the bucket.

4. The drop sill length L should not be less than 1.5 times the contracted water depth h_1 on the bucket.

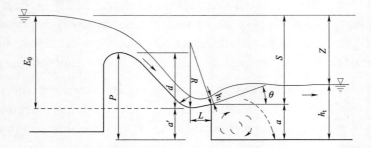

Key

E_0 upstream total head

P height of weir above the downstream riverbed

d height from the lowest point of reverse arc to weir crest

a' height from the downstream riverbed to the lowest point of reverse arc

R reverse arc radius

h_1 contracted water depth on the bucket

L drop sill length from the lowest point of reverse arc to the bucket lip

θ trajectory angle

S elevation difference from bucket top to upstream water surface

a bucket lip height

Z difference between upstream and downstream water level

h_t water depth of the riverbed downstream

Figure 7.2.2 Configuration parameters of surface flow energy dissipation

7.2.3 For the surface flow energy dissipation design, the hydraulic calculation under various flows and possible corresponding water level combinations shall be conducted to ensure the smooth transition from free surface flow to submerged surface flow under various flows.

7.2.4 The configuration of bucket-type flow energy dissipation may be of solid or slotted type. The configuration design parameters such as bucket trajectory angle θ, reverse arc radius R, bucket lip height a and invert elevation should be determined by hydraulic model tests. The configuration parameters of solid bucket (Figure 7.2.4) should be taken within the following empirical ranges:

1 The bucket trajectory angle θ may be 30° to 45°.

2 The reverse arc radius R may be selected according to the empirical range of the ratio of upstream total head E_0 above the bucket invert to the reverse arc radius R. The empirical ratio of E_0 to R should be 2.1 to 8.4.

3 The bucket lip height a should be determined together with the bucket trajectory angle θ, reverse arc radius R and the height Δb of bucket invert above downstream riverbed, and should be taken as 1/6 of the water depth of the riverbed downstream h_t.

4 The elevation of bucket invert may be higher than or equal to that of the downstream riverbed.

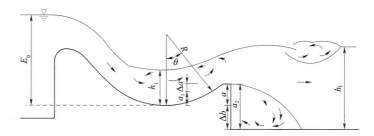

Key

E_0 upstream total head above the bucket invert
h_1 contracted flow depth at the bucket invert
θ bucket trajectory angle
R bucket radius
Δa height difference from the top of bucket lip to the end point of arc
a_1 height difference from the bucket invert to the end point of arc
a height of bucket lip
Δb height difference from riverbed to the bucket invert
a_2 height difference from riverbed to top of bucket lip
h_t water depth of the riverbed downstream

Figure 7.2.4 Configuration parameters of solid bucket

7.2.5 For the bucket-type flow energy dissipation design, the hydraulic calculation under various flows and possible corresponding water level combinations shall be conducted to ensure favorable bucket flow patterns under various discharge flows.

7.2.6 When a stilling basin for bucket-type energy dissipation works together with end-flared piers, its configuration and hydraulic design shall be comprehensively determined by topographical and geological conditions, layout of adjacent structures, river changes, downstream protection, and shall meet the following requirements:

1 The horizontal length of the stilling basin may be preliminarily determined empirically, ensuring that the outer edge of nappe behind the end-flared pier falls into the basin when releasing the check flood.

2 The gradient range of end sill of the stilling basin is 1 : 1.5 to 1 : 2.5. The height of end sill may be taken as 1/5 of the design downstream water depth.

3 When the end sill needs to be heightened because of some special requirements such as maintenance, it shall be demonstrated through hydraulic model tests.

7.3 Structural Design

7.3.1 The type and size of surface flow and bucket-type flow energy dissipation structures shall be determined according to the layout, hydraulic design, foundation, and operating conditions, taking into account engineering measures such as seepage control, drainage, waterstop and anchorage, and the design content shall include:

1 Structural type selection and layout.

2 Calculation of stability against sliding and floating.

3 Calculation of structural stresses.

4 Concrete strength, frost resistance, impermeability, and temperature control requirements.

7.3.2 The design of resistance against abrasion and cavitation for surface flow and bucket-type flow energy dissipation structures shall meet the requirements of Article 3.0.6 of this guide.

8 Design for Inside-Tunnel Energy Dissipation and Erosion Control

8.1 General Requirements

8.1.1 The layout of inside-tunnel energy dissipation shall be determined through comprehensive techno-economic comparison according to the topographical and geological conditions, project layout conditions, renovation and utilization conditions, etc.

8.1.2 The intake may adopt open or deep-seated inlets. On the premise of a reasonable inlet layout and smooth water flow, the energy dissipation tunnel section shall be arranged in intact rock mass.

8.1.3 When the inside-tunnel energy dissipation is adopted, the discharge capacity, flow pattern, energy dissipation characteristics, time-averaged pressure, fluctuation pressure and cavitation characteristics of the tunnel under various operating conditions shall be analyzed through hydraulic model tests, so as to reasonably determine the configuration of the energy dissipator in the tunnel. The hydraulic calculation for inside-tunnel energy dissipation should be in accordance with Appendix F of this guide.

8.1.4 In addition to NB/T 10391, *Code for Design of Hydraulic Tunnel*, the structural design for inside-tunnel energy dissipation shall also meet the requirements of Article 3.0.6 of this guide.

8.2 Configuration Design

8.2.1 The shaft swirl flood discharge tunnel may consist of water inlet, approach, swirling chamber, vertical shaft, stilling shaft, ventilation shaft, release tunnel, and outlet section. For the shaft spillway with swirling flow, the configuration design of energy dissipator shall meet the following requirements.

1. The approach may be of free flow or pressure flow. The outer sidewall of approach and the sidewall of swirling chamber shall be connected tangentially by a straight line or eccentric elliptic curve. Guide sills should be set for water flowing from approach into the swirling chamber.

2. The diameter of swirl shaft may be estimated by the method given in Section F.1 of this guide.

3. The cross-section of the swirling chamber should be in single arc shape, and the diameter may be taken as 1.2 to 1.6 times the vertical shaft diameter.

4 The swirling chamber and shaft should be connected by a conical transition with a contraction angle not greater than 11.5° and a length greater than 1.0 times the shaft diameter.

5 A stilling shaft may be set at the lower part of the shaft, and its diameter may be equivalent to that of the shaft, and the depth may be 1.0 to 1.2 times the diameter of the shaft.

6 A contraction-pressing plate should be set at the top of the connection between the release tunnel and the shaft. The height of the orifice at the outlet of the contraction-pressing plate may be so designed that the flow velocity at the outlet equals the average velocity of uniform flow in the release tunnel.

7 A ventilation shaft should be set at the top of the closed swirling chamber and behind the contraction-pressing plate.

8.2.2 The horizontal vortex flood discharge tunnel may consist of water inlet, approach, vertical shaft, ventilation shaft, swirling chamber, swirling tunnel, water cushion chamber, baffle block, and release tunnel. Its configuration design shall meet the following requirements:

1 The diameter of the horizontal vortex tunnel may be estimated by the method given in Section F.1 of this guide.

2 The cross-sectional diameter of the swirling chamber may be 1.1 to 1.2 times the diameter of the horizontal vortex tunnel, and the eccentricity between the centerline of the swirling chamber and that of the vertical shaft may be 0.32 to 0.50 times the diameter of the horizontal vortex tunnel. The size of the contracted section at the swirling chamber outlet shall be coordinated with that of water inlet and shaft to meet the discharge capacity requirements of the tunnel.

3 Baffle blocks may be set at a certain distance behind the horizontal vortex tunnel to form a water cushion chamber between the horizontal vortex tunnel and the baffle blocks. The horizontal vortex tunnel and the water cushion chamber may be connected by a transition section.

4 A ventilation shaft should be set upstream of the horizontal vortex tunnel.

8.2.3 The orifice plate flood discharge tunnel may consist of water inlet, pressure connection section, energy dissipation section with orifice plates, and outlet section. The configuration design of the orifice plate energy dissipator should meet the following requirements:

1. The pressure connection section should be set upstream of the impervious curtain line of the dam.

2. The ratio of orifice plate spacing to the tunnel diameter may be taken as 4.5. The hole edge of each orifice plate is in the shape of circular arc with different radius, and the aperture ratio should not be greater than 0.75.

3. The gate control section should be arranged downstream of the energy dissipation section with multi-stage orifice plant to avoid the occurrence of transient regime of alternating free flow and pressure flow in the tunnel when the gate is opened.

8.2.4 The thick orifices flood discharge tunnel may consist of water inlet, pressure connection section, energy dissipation section with thick orifice, and outlet section. The configuration design of thick orifice energy dissipator should meet the following requirements:

1. The pressure connection section and energy dissipation section with thick orifice should be set upstream of the impervious curtain of the dam.

2. The thick orifice should adopt circular cross-section. The spacing between thick orifices should be greater than 3.0 times the tunnel diameter. The ratio of the sectional area of the thick orifice to that of the tunnel may be taken as 0.2 to 0.5. The flow-passing section size of the downstream thick orifice should be larger than that of the upstream one.

8.3 Hydraulic Design

8.3.1 The hydraulic design for inside-tunnel energy dissipation shall meet the requirements of discharge capacities under various operating conditions, and the impact of cavitation, scouring, vibration and dynamic load on the safe operation of the project shall be analyzed.

8.3.2 The hydraulic design of the shaft swirl energy dissipator of flood discharge tunnel shall meet the following requirements:

1. The flow pattern in the tunnel section of the approach shall be stable, the vortex shall be formed in the swirling chamber, and a stable cavity shall be maintained in the center of the swirling chamber and the shaft. The stilling shaft and contraction-pressing plates shall be able to dissipate and regulate the flow pattern, and the water flow in the release

tunnel shall be smooth.

2 No negative pressure should appear on the flow-passing surface of the tunnel with shaft swirl energy dissipator.

8.3.3 The hydraulic design of the horizontal vortex energy dissipator of flood discharge tunnel shall meet the following requirements:

1 The shaft section should be of pressure flow, and the average velocity should not be greater than 20 m/s.

2 Effective vortex shall be formed in the swirling chamber, and a stable cavity shall be formed in the center of the swirling tunnel.

3 No negative pressure shall appear on the flow-passing surface of the tunnel with horizontal vortex energy dissipator.

8.3.4 The hydraulic design of the orifice plate energy dissipator of flood discharge tunnel shall meet the following requirements:

1 The energy dissipation section with orifice plates shall be of pressure flow, and the transient regime of alternating free flow and pressure flow shall be avoided during operation.

2 Measures such as optimal design of geometric size, use of abrasion-resistant and cavitation-resistant materials, and reasonable operation mode shall be taken to avoid cavitation erosion.

3 The fluctuation pressure distribution, frequency and intensity of turbulent flow induced by orifice plate energy dissipation shall be analyzed to ensure the safe operation of the tunnel.

4 The head loss of orifice plates may be estimated by the method given in Section F.2 of this guide.

8.3.5 The hydraulic design of the thick orifice energy dissipator of flood discharge tunnel shall meet the following requirements:

1 The flow velocity in the pressure connection section should not exceed 20 m/s, and the thick orifice section shall be of pressure flow.

2 The minimum pressure at the crown of the pressure connection section should not be less than 0.02 MPa under the most unfavorable operating condition. The pressure at the tunnel crown may be calculated by the energy equation method.

3 Reasonable geometric size shall be selected for the thick orifice section, and measures against cavitation erosion, such as surface protection

materials with good corrosion resistance and reasonable operation mode, shall be adopted.

4   The head loss of thick orifice may be estimated by the method given in Section F.3 of this guide.

9 Downstream Protection Design

9.1 General Requirements

9.1.1 The design of downstream protection works shall meet the following requirements:

1. The downstream protection works shall be determined through comprehensive techno-economic comparison according to the protected objects and corresponding protection criteria, topographical and geological conditions, energy dissipation types, navigation requirements, construction and operation conditions, etc.

2. The downstream protection works shall be coordinated with the layout of power generation, navigation and floating debris drifting to avoid mutual interference, and shall also highlight the harmonization with the surroundings and landscaping.

3. The object to be protected shall be safe and reliable itself. If the protected object has safety problems such as instability, it shall be treated and then protected or may be treated together with protective measures which shall not have an adverse impact on the protected object.

4. For the protection of existing buildings or structures of electric power, land transportation, navigation, water resources, and industrial and civil buildings, and other facilities in the protection scope, the protection scheme shall be approved by the authorities, and special analysis shall be conducted when necessary.

5. The stability of downstream bank slopes shall comply with the current sector standard DL/T 5353, *Design Specification for Slope of Hydropower and Water Conservancy Project*.

9.1.2 The scope of downstream protection works shall meet the following requirements:

1. The bank rock slope that meets the scouring requirements and is stable, and the riverbed where the extent and depth of scour hole do not endanger the safety of main structures or slopes, need not be protected.

2. The protection scope of hydraulic jump energy dissipation, surface flow energy dissipation and bucket-type flow energy dissipation shall include the stilling basin and the riverbed and bank slopes downstream of the stilling bucket. The protection scope of ski-jump energy

dissipation shall include the riverbed, bank slopes and atomization-affected area downstream of the flip bucket if there is no plunge pool, and shall include the bank slopes upstream of the plunge pool and the riverbed, bank slopes and atomization-affected area downstream of the plunge pool if there is a plunge pool. The protection scope of inside-tunnel energy dissipation such as swirl and orifice plate shall include the riverbed and bank slopes downstream of the tunnel outlet.

3   The protection height of downstream bank slopes shall be determined according to the water level corresponding to the design flood standard of energy dissipation and erosion control structures.

4   The scouring range of downstream riverbed and bank slopes should be determined through hydraulic model tests. In the absence of test data, the scour calculation of downstream river channel may be conducted in accordance with Appendix G of this guide.

5   When the anti-scouring capability of slopes in the flood discharge atomization area is not sufficient, or the slope stability may fail because of the rainstorm caused by atomization, the slopes in the atomization area shall be protected.

9.1.3   The design of downstream protection works shall include the following:

1   Collect and analyze the basic data for the protection works such as hydrology, meteorology, topography and geology, and the distribution of downstream buildings or structures.

2   Analyze and determine the scouring degree and scope of downstream riverbed and bank slopes, and propose the safety criteria and anti-scouring criteria for the protection design of protected objects.

3   Analyze the hydraulic parameters of the downstream river channel under the design flood and below, analyze the scope and type of downstream protection, conduct the design of protection structures and foundation treatment, and propose engineering measures such as slope reinforcement and drainage, according to the project layout and the energy dissipation mode.

4   Study and propose protective measures for other buildings or structures in the protection scope.

5   Predict and analyze the flood discharge atomization range, and determine the atomization zones by rainfall intensity. Analyze the possible hazards and influence degree of flood discharge atomization,

and propose protection and drainage measures accordingly.

9.1.4 The downstream protection structures shall meet the following requirements:

1. Under the design flood and below for energy dissipation and erosion control, the protection structures shall meet the requirements of protection function and self-safety.

2. Under the flood exceeding the design flood of energy dissipation and erosion control, the local damage of downstream protection structure shall not affect the normal operation of the main structures, and the repairing conditions of downstream protection structure shall be available.

9.2 Protection Type and Structural Design

9.2.1 The downstream protection type shall be selected considering the hydraulic characteristics of flood discharge, topographical and geological conditions of river channel, and stability conditions of slopes in the flood discharge atomization area. The protection type shall be determined according to the following conditions:

1. Hydraulic factors such as velocity, wave, water level, and scouring and silting form of the discharged flow in the downstream river channel.

2. The allowable non-scouring velocity and anti-scouring capacity of the downstream river channel shall be determined according to the engineering geological conditions of the river channel. In the absence of relevant data, the empirical allowable non-scouring velocity for downstream channel may be taken in accordance with Appendix H of this guide.

3. Water table at downstream bank slope.

9.2.2 The protection type of downstream bank slopes shall be determined according to the stability of the bank slopes, geological and topographical conditions, weathering degree, etc., and should meet the following requirements:

1. When the flow velocity and wave in the river are small, flexible slope protection such as dry masonry, gabions, protective net, or shotcrete anchor support should be adopted.

2. When the flow velocity or wave in the river is large, rigid slope protection such as concrete and stone masonry should be adopted. The

protection should be of retaining wall, slope paving, their combination, etc.

3   When the bank slope is unstable, the protection type should be determined considering the bank slope stability control measures.

9.2.3   The protection type of downstream riverbed shall be determined according to the riverbed geology, flow velocity, etc., and may be of apron, apron extension, etc., which should meet the following requirements:

1   When the flow velocity is small, flexible bottom protection such as dry masonry, riprap and gabions should be adopted.

2   When the flow velocity is large, rigid bottom protection such as concrete and masonry with pressure-balancing holes should be adopted.

9.2.4   The protection types, such as anti-scouring wall, fascine mattress, spur dike, longitudinal dike, and submerged dike, may be adopted when it is difficult to solve the scouring of downstream bank slopes and riverbed by adopting the protection types given in Articles 9.2.2 and 9.2.3 of this guide, or when it is necessary to accommodate the special flow patterns such as downstream navigation and ecology.

9.2.5   The protection structure and protected object shall be reliably combined. The protected sections of bank slope, the protected section and the unprotected section, and the bottom protection structure and the downstream riverbed shall be smoothly connected. Drainage system should be set for slope protection.

9.2.6   The protection structure shall be determined according to its materials, foundation geological conditions and load analysis, and the structural calculation shall meet the following requirements:

1   Structural stability and stress calculation shall be carried out for slope protection and retaining wall.

2   The structural design of longitudinal dike, spur dike and submerged dike may comply with the current national standard GB 50286, *Code for Design of Levee Project*.

9.2.7   The partitioning of protection structures shall be comprehensively determined according to foundation geological conditions, materials, structural stress conditions, etc.

9.2.8   The protection structures may adopt the cut-off trench, expanded foundation, pile foundation, etc. For rock foundation, dowel bars should be set, and consolidation grouting may be adopted for the foundation.

9.3 Protection of Flood Discharge Atomization Zone

9.3.1 The flood standard for flood discharge atomization area shall be the same as that of energy dissipation and erosion control structure of the project. The flood discharge atomization range and atomization rainfall intensity may be determined with reference to similar projects. Numerical simulation or hydraulic model tests should be conducted for important projects and projects suffering greatly from flood discharge atomization.

9.3.2 The flood discharge atomization area may be zoned into water nappe splitting and splashing area, dense fog rainstorm area, mist rainfall area and light fog water vapor dispersion area, and may be protected according to the affecting range and distribution of atomization rainfall intensity, taking into account the local natural rainfall intensity and frequency.

9.3.3 Switchyards, power station outlets, high-voltage lines, surface powerhouse, traffic facilities, etc. should not be arranged in water nappe splitting and splashing area or dense fog and rainstorm area. Effective protective measures shall be taken when inevitable.

9.3.4 The protective measures for slopes in the flood discharge atomization area shall be determined comprehensively according to the geological conditions, slope grade, and slope stability analysis, taking into account the protective performance, difficulty in construction, project investment, etc. The atomization rainfall intensity zoning and protective measures may be selected in accordance with Table 9.3.4.

Table 9.3.4 Atomization rainfall intensity zoning and protective measures

S.N.	Zone	Atomization rainfall intensity S (mm/h)	Protective measures
1	Water nappe splitting and splashing area	$S \geq 50$	Concrete slope protection, drainage system
2	Dense fog rainstorm area	$50 > S \geq 10$	Concrete slope protection, shotcrete slope protection, drainage system
3	Mist rainfall area	$10 > S \geq 2.0$	Need not be protected
4	Light fog water vapor dispersion area	$S < 2.0$	No protection

10 Hydraulic Model Test and Hydraulic Numerical Simulation

10.1 General Requirements

10.1.1 When analyzing the correlation of the discharge flow of water release structure with the related structures, the scouring of the downstream river channel and the hydraulic effect of the bank slopes, integral hydraulic model tests should be conducted.

10.1.2 When analyzing the hydraulic characteristics of water release structure, cross-section or local model tests should be conducted.

10.1.3 The energy dissipation and erosion control structures with novel energy dissipators, abrupt changes in boundary or complex hydraulic characteristics shall be demonstrated by cross-section or local model tests.

10.1.4 Depressurization model tests or flow-induced vibration model tests shall be conducted for energy dissipators subjected to potential damage by cavitation erosion or vibration.

10.1.5 Hydraulic numerical calculation may be used in the analysis of energy dissipation and erosion control scheme.

10.2 Hydraulic Model Test

10.2.1 The principle and data analysis of hydraulic model tests shall comply with the current sector standards DL/T 5244, *Code for Normal Hydraulics Model Investigation for Hydropower & Water Resources*; and DL/T 5245, *Code for Hydraulic Investigation on Aeration-Cavitation Resistance for Hydropower & Water Resources*.

10.2.2 The content and requirements of hydraulic model tests shall be determined according to the scale of water release structures and energy dissipation and erosion control structures. Appropriate similarity criteria and model scale shall be chosen.

10.2.3 The hydraulic model tests shall cover the discharge capacity, water surface profile, flow velocity, flow pattern, time-averaged hydrodynamic pressure, scouring and silting of downstream riverbed, wave, fluctuation pressure and velocity near the bottom of water release structures, and starting ski-jump discharge.

10.2.4 According to the project characteristics, the hydraulic phenomena occurring during the change of operating conditions should be observed and described when conducting hydraulic model tests.

10.2.5 After the determination of flood discharge scheme, flood regulation test should be conducted to propose the operation mode of water release structures.

10.2.6 The hydraulic model test for erosion control of downstream river channel shall mainly include the water surface profile and waves near both banks, the velocity distribution along the river bank, and the scouring and silting distribution of the channel.

10.2.7 The flow-induced vibration test shall focus on the flow pressure fluctuation under various operating conditions, structural vibration mode and corresponding parameters, and dynamic response parameters of flow-induced vibration of structures under various operating conditions. The flow-induced vibration model test shall comply with the current sector standard SL 158, *Code for Model Test on Flow Pressure Fluctuation and Flow Induced Vibration of Hydraulic Structure*.

10.2.8 The depressurization model test shall comply with the current sector standard DL/T 5359, *Code for Model Test of Cavitation for Hydropower & Hydraulic Engineering*.

10.2.9 The aeration-cavitation test shall comply with the current sector standard DL/T 5245, *Code for Hydraulic Investigation on Aeration-Cavitation Resistance for Hydropower & Water Resources*.

10.3 Hydraulic Numerical Simulation

10.3.1 The software used for hydraulic numerical calculation shall be well-proven and reliable, and the parameters used in the mathematical model should be verified.

10.3.2 The scope of hydraulic numerical calculation shall include water release structures, and upstream and downstream areas to ensure a stable flow pattern.

10.3.3 The hydraulic numerical calculation shall include the discharge capacity, flow pattern, pressure, water level, flow velocity, and flow-induced vibration.

10.3.4 A well-proven model shall be used in the numerical calculation of flood discharge atomization, and should be verified for reliability before calculation based on the prototype observation data of similar projects. The results shall cover the water spray concentration and the rainfall intensity isograph.

11 Safety Monitoring Design

11.1 General Requirements

11.1.1 The safety monitoring of energy dissipation and erosion control structures may include structural safety monitoring and hydraulic prototype observation.

11.1.2 For the energy dissipation and erosion control structures with complex hydraulic conditions in a large hydropower project, hydraulic prototype observation design shall be performed.

11.1.3 The items of structural safety monitoring and hydraulic prototype observation of energy dissipation and erosion control structures shall be determined according to the structural and hydraulic characteristics of water release structures.

11.2 Structural Safety Monitoring

11.2.1 The monitoring objects shall be determined by the type of energy dissipation and erosion control facility of a water release structure. The monitoring objects should be the foundation and floor for hydraulic jump energy dissipation, plunge pool and end dam for ski-jump energy dissipation, bucket and drop sill for free surface flow energy dissipation, and surrounding rock and lining structure for inside-tunnel energy dissipation. For a novel energy dissipator, the monitoring objects shall be determined according to its structural characteristics.

11.2.2 The structural safety monitoring items of Grades 1, 2 and 3 water release structures shall be selected from Table 11.2.2. Grades 4 and 5 water release structures need not undergo structural safety monitoring unless otherwise specified.

Table 11.2.2 Structural safety monitoring items

Monitoring item	Grade of water release structure		
	1	2	3
Foundation uplift pressure	★	★	×
Stresses of anchor bolt/cable	★	△	×
Joint openness	★	△	△
Seepage flow	★	△	×
Reinforcement stresses	★	△	×
Surrounding rock deformation	★	★	△
Vibration	△	△	×
Erosion	△	△	×

NOTE ★ denotes mandatory; △ denotes optional; × denotes unnecessary.

11.2.3 The key monitoring parts shall be those with large flow velocity, large fluctuation pressure, or poor geological conditions. Uplift pressure should be monitored for the floor of stilling basin or plunge pool. Joint openness should be monitored for the joints between stilling basin floor and bedrock, floor structural joints, joints between tunnel lining and surrounding rock, etc. Reinforcement stresses should be monitored for the reinforcement of energy dissipator floor or surrounding rock lining. Stresses should be monitored for the floor of energy dissipator or tunnel surrounding rock bolts and anchor bundles.

11.2.4 The real-time monitoring scheme of energy dissipation and erosion control structures during flood discharge should be proposed.

11.3 Hydraulic Prototype Observation

11.3.1 The hydraulic prototype observation design of energy dissipation and erosion control structures shall meet the following requirements:

1. The observation items shall reflect the main hydraulic characteristics of energy dissipation and erosion control structures. The optional items include flow pattern, hydrodynamic pressure, flow velocity, water level and wave, cavitation noise, aeration concentration, vibration, local scouring and silting, flood discharge atomization, wind speed during flood discharge, flood discharge noise, and environmental quantity.

2. The observation points should correspond to the key points of hydraulic model tests, and shall be arranged according to actual needs for special energy dissipation mode.

3. For energy dissipation and erosion control structures with hydraulic jump, surface flow, and bucket-type flow energy dissipation, hydraulic observation points should be set in and around the hydraulic jump area.

4. For energy dissipation and erosion control structures with ski-jump and drop energy dissipation, hydraulic observation points should be set in a certain area of flip bucket and nappe impingement.

5. The key observation items shall include flow pattern, fluctuation pressure of energy dissipation and erosion control structures, plate vibration, flow velocity and other hydraulic characteristics near bottom, local scouring of plunge pool with bank slope protection but without bottom protection, and structural stability of guide wall.

6. The observation items for cavitation noise should be set in the shape change area of high velocity flow surface and the hydraulic jump of energy dissipator.

11.3.2 The hydraulic prototype observation items of downstream protection works should include flow pattern, near-bank flow velocity, water level, local scouring, and wave.

11.3.3 The hydraulic prototype observation design of flood discharge-affected area shall meet the following requirements:

1. The hydraulic prototype observation station shall be set in a safe area of the energy dissipator.

2. Observation points for atomization should be arranged behind the dam and on the bank slopes, roads, switchyard, outlet of high-voltage power lines, powerhouse and other parts affected by atomization.

3. Ecological environment quantities should be observed at the gate operation room, major roads, residential area and office area before and during the flood discharge. The observation parameters may include flood discharge noise, wind speed during flood discharge, vibration, air fluctuation, temperature, humidity, etc.

12 Operation and Maintenance Requirements

12.0.1 According to the reservoir operation requirements, technical requirements for the operation mode of water release structures, operation of gates, and operation and maintenance of energy dissipation and erosion control structures shall be proposed, considering the hydraulic model tests, hydraulic prototype observation and engineering analogy.

12.0.2 Energy dissipation and erosion control structures shall be inspected and maintained before and after each flood season.

Appendix A Hydraulic Calculation for Hydraulic Jump Energy Dissipation

A.0.1 The hydraulic jump on a horizontal smooth apron (Figure A.0.1) shall meet the following requirements:

1. The subsequent conjugate depth of free hydraulic jump can be calculated by the following formulae:

$$h_2 = \frac{h_1}{2}\left(\sqrt{1+8Fr_1^2} - 1\right) \tag{A.0.1-1}$$

$$E_0 = h_1 + \frac{q^2}{2g\varphi_0^2 h_1^2} \tag{A.0.1-2}$$

$$Fr_1 = \frac{v_1}{\sqrt{gh_1}} \tag{A.0.1-3}$$

where

h_2 is the conjugate depth (m);

h_1 is the pre-jump water depth at contracted section (m);

Fr_1 is the pre-jump Froude number of contracted section;

E_0 is the total head above the bottom of the contracted section (m);

q is the unit discharge [m³/(s · m)];

g is the acceleration of gravity (m/s²);

φ_0 is the velocity coefficient of contracted section, which may be taken as 0.95 in estimate;

v_1 is the pre-jump average velocity of contracted section (m/s).

2. The judgement of hydraulic jump form shall meet the following requirements:

  1) When the conjugate depth is obviously larger than downstream water depth and hydraulic jump cannot be formed, it is rapid flow.

  2) When the conjugate depth is slightly larger than downstream water depth, it is remote hydraulic jump.

  3) When the conjugate depth is equal to downstream water depth, it is critical hydraulic jump.

  4) When the conjugate depth is smaller than downstream water depth, it is submerged hydraulic jump.

3 The free hydraulic jump length can be calculated by the following formula:

$$L = 9.4(Fr_1 - 1)h_1 \qquad (A.0.1\text{-}4)$$

where

L is the length of free hydraulic jump (m).

4 When Fr_1 is 4.5 to 15.5, the free hydraulic jump length can also be estimated as follows:

$$L = (5.9 \sim 6.15)h_2 \qquad (A.0.1\text{-}5)$$

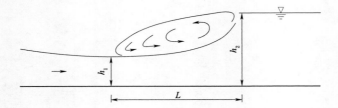

Key

h_1 pre-jump water depth at contracted section

L free jump length

h_2 conjugate depth

Figure A.0.1 Hydraulic jump on horizontal smooth apron

A.0.2 The hydraulic jump in down-cut stilling basin with equal-width rectangular section (Figure A.0.2) may be calculated as follows:

1 The depth of stilling basin can be calculated by the following formulae:

$$S = \sigma h_2 - h_t - \Delta Z \qquad (A.0.2\text{-}1)$$

$$\Delta Z = \frac{q^2}{2g}\left(\frac{1}{\varphi_1^2 h_t^2} - \frac{1}{\sigma^2 h_2^2}\right) \qquad (A.0.2\text{-}2)$$

where

S is the depth of stilling basin (m);

σ is the submergence degree of hydraulic jump, which may be taken as 1.05 to 1.10;

h_t is the water depth on downstream riverbed (m);

ΔZ is the drop at the basin end sill (m);

φ_1 is the velocity coefficient of stilling basin outflow, which may be taken as 0.95.

2 The length of stilling basin can be estimated by the following formula:

$$L_k \approx 0.8 L \tag{A.0.2-3}$$

where

    L_k is the length of stilling basin (m).

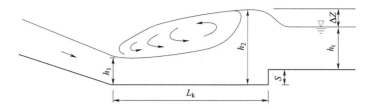

Key

h_1 pre-jump water depth at contracted section
L_k length of stilling basin
h_2 conjugate depth
S depth of stilling basin
ΔZ drop at the basin end sill
h_t water depth on downstream riverbed

Figure A.0.2 Hydraulic jump in down-cut stilling basin with equal-width rectangular section

A.0.3 The hydraulic jump in end sill stilling basin with equal-width rectangular section (Figure A.0.3) may be calculated as follows:

1 The height of end sill can be calculated by the following formulae:

$$C = \sigma h_2 - H_1 \tag{A.0.3-1}$$

$$H_1 = H_{10} - \frac{q^2}{2g(\sigma h_2)^2} \tag{A.0.3-2}$$

$$H_{10} = \left(\frac{q}{m_1\sqrt{2g}}\right)^{2/3} \tag{A.0.3-3}$$

$$H_1 = \left(\frac{q}{\sigma_s m_1 \sqrt{2g}}\right)^{2/3} - \frac{q^2}{2g(\sigma h_2)^2} \tag{A.0.3-4}$$

where

    C is the height of end sill (m);

    H_1 is the overflow depth on end sill top (m), determined by the outflow submergence, which shall be calculated by Formula (A.0.3-2) for free outflow, or by Formula (A.0.3-4) for submerged outflow;

H_{10} is the total head of overflow at end sill top (m);

m_1 is the flow coefficient for stilling basin outflow, which may be taken as 0.42 in estimate;

σ_s is the submergence coefficient for end sill outflow. If $h_s/H_{10} \leq 0.45$, the end sill is a non-submerged weir, and the submergence coefficient is 1; if $h_s/H_{10} > 0.45$, it is a submerged weir. The submergence coefficient may be in accordance with Table A.0.3.

Table A.0.3 Submergence coefficient for stilling basin end sill

h_s/H_{10}	≤0.45	0.50	0.55	0.60	0.65	0.70	0.72	0.74	0.76	0.78
δ_s	1.000	0.990	0.985	0.975	0.960	0.940	0.930	0.915	0.900	0.885
h_s/H_{10}	0.80	0.82	0.84	0.86	0.88	0.90	0.92	0.95	1.00	–
δ_s	0.865	0.845	0.815	0.785	0.750	0.710	0.651	0.535	0.000	–

2 The length of stilling basin can be estimated by Formula (A.0.2-3).

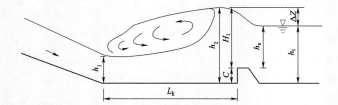

Key

h_1 pre-jump depth at contracted section
L_k length of stilling basin
h_2 conjugate depth
H_1 overflow depth on end sill top
C height of end sill
h_s water depth from the downstream water surface to the end sill top
ΔZ drop at the basin end sill
h_t water depth on downstream riverbed

Figure A.0.3 Hydraulic jump in end sill stilling basin with equal-width rectangular section

A.0.4 Hydraulic jump in combined stilling basin with equal-width rectangular section (Figure A.0.4) may be calculated as follows:

1 The end sill height of combined stilling basin can be calculated by the

following formulae:

$$C = h_{t1} + \frac{q^2}{2g\varphi_1^2 h_{t1}^2} - H_{10} \quad \text{(A.0.4-1)}$$

$$h_{t1} = \frac{h_t}{2}\left(\sqrt{1 + 8\frac{q^2}{gh_t^3}} - 1\right) \quad \text{(A.0.4-2)}$$

where

h_{t1} is the water depth of contracted section in the case of secondary hydraulic jump behind the end sill (m).

2 The down-cut depth of the combined stilling basin can be calculated by the following formulae:

$$S = \sigma h_2 - C - H_1 \quad \text{(A.0.4-3)}$$

$$H_1 = \left(\frac{q}{m_1\sqrt{2g}}\right)^{2/3} - \frac{q^2}{2g(\sigma h_2)^2} \quad \text{(A.0.4-4)}$$

where

H_1 is the overflow water depth on the end sill top in the case of free outflow (m).

3 The length of stilling basin can be estimated by Formula (A.0.2 - 3).

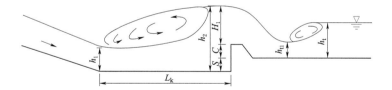

Key

- h_1 pre-jump depth at contracted section
- L_k length of stilling basin
- h_2 conjugate depth
- H_1 overflow depth on end sill top
- C height of end sill
- S down-cut depth of stilling basin
- h_{t1} water depth of contracted section in the case of secondary hydraulic jump behind the end sill
- h_t water depth on downstream riverbed

Figure A.0.4 Hydraulic jump in combined stilling basin with equal-width rectangular section

A.0.5 The conjugate depth ratio of hydraulic jump in sloping channels (Figure A.0.5-1) can be calculated by the following formula:

$$\lambda = \frac{h_2}{h_1} = \frac{1}{\cos\theta}\left(\sqrt{\frac{2Fr_1^2 \cos^3\theta}{1-2C'\tan\theta}+0.25}-0.5\right) \tag{A.0.5}$$

where

    λ conjugate depth ratio of hydraulic jump in sloping channel;

    θ angle between sloping channel and horizontal line (°);

    C' correction factor, which may consult Figure A.0.5-2.

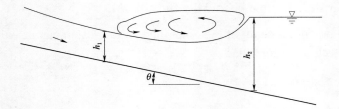

Key

h_1 pre-jump depth at contracted section

θ angle between sloping channel and horizontal line

h_2 conjugate depth

Figure A.0.5-1   Hydraulic jump in sloping channel

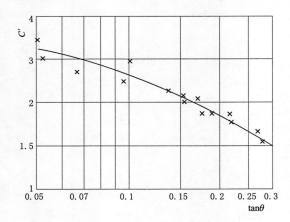

Key

×   experimental data

C' correction factor

θ angle between sloping channel and horizontal line

Figure A.0.5-2   Relationship between C' and $\tan\theta$

Appendix B Configuration Parameters for End-Flared Pier

B.0.1 The main hydraulic parameters of end-flared pier (Figure B.0.1) can be preliminarily proposed by the following formulae:

$$\frac{h_d}{P_d} = 23.94K^3 - 20.4K^2 + 6.5K + 0.0123 \qquad (B.0.1-1)$$

$$K = q/\sqrt{g}Z^{1.5} \qquad (B.0.1-2)$$

where

h_d is the elevation difference between downstream water level and stilling basin floor (m);

P_d is the elevation difference between weir crest and stilling basin floor (m);

h_d/P_d is the dimensionless parameter;

K is the flow energy ratio representing the dynamic characteristics of dam discharge, which is a dimensionless parameter;

q is the unit discharge [m³/(s · m)];

g is the acceleration of gravity (m/s²);

Z is the difference between upstream and downstream water levels (m).

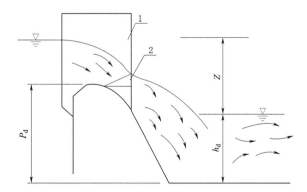

Key

1   gate pier

2   end-flared pier

P_d elevation difference between weir crest and stilling basin floor

Z difference between upstream and downstream water levels

h_d elevation difference between downstream water level and stilling basin floor

Figure B.0.1 Main hydraulic parameters of end-flared pier

B.0.2 The selection of configuration parameters of Y-shaped end-flared pier (Figure B.0.2-1) shall meet the following requirements:

1   Contraction ratio and lateral contraction angle can be preliminarily proposed by the following formulae:

$$\varepsilon = 0.2 \frac{\eta}{\tan\theta} - 0.06 \tag{B.0.2-1}$$

$$\eta = \frac{H_d}{Z_a} \tag{B.0.2-2}$$

$$\tan\theta = \frac{B - b_0}{2L} \tag{B.0.2-3}$$

where

ε is the contraction ratio, which should be taken as 0.333 to 0.667, and may consult Figure B.0.2-2 for the $\varepsilon - \eta / \tan\theta$ relation;

η is the bottom position parameter of end-flared pier end section, which may be taken as 0.38 to 0.92;

θ is the lateral contraction angle, which should be taken as 12° to 22°;

H_d is the weir crest design head (m);

Z_a is the elevation difference between the weir crest and the lowest point of end-flared pier end section (m);

B is the clear distance between piers (m);

b_0 is the clear distance between end sections of end-flared piers (m);

L is the longitudinal horizontal projection length of end-flared pier (m).

2   After the boundary conditions such as the size of the gate orifice and the radius of the radial gate are determined, Formula (B.0.2-1) may be used to initially determine the main dimensions of end-flared pier under the weir crest design head H_d.

3   For an end-flared pier, point B and point C should be equal in elevation, or point C should be slightly lower than point B. The height h of end section CD of end-flared pier should be 0.6 times Z_a, the elevation difference between the weir crest and the lowest point of end section CD. When the downstream water level is higher than point D, the height h should not be less than 0.7 times Z_a, the elevation difference between the weir crest and the lowest point of end section CD.

4   When proposing the size of end-flared pier, the sizes of end-flared piers of side-orifice shall be appropriately adjusted according to the

layout and hydraulic conditions. The configuration parameters shall be finalized by hydraulic model tests.

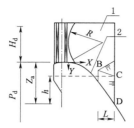
(a) Elevation view of Y-shaped end-flared pier

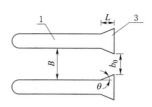
(b) Plan view of Y-shaped end-flared pier

Key

1   gate pier
2   WES curve
3   end-flared pier
B   upper break point of the end-flared pier
C   upper contraction point of the end-flared pier
D   lower contraction point of the end-flared pier
H_d weir crest design head
P_d elevation difference between weir crest and stilling basin floor
Z_a elevation difference between the weir crest and the lowest point of end-flared pier end section
h height of end-flared pier end section CD
R radius of radial gate
B clear distance between piers
θ lateral contraction angle
b_0 clear distance between end sections of end-flared piers

Figure B.0.2-1   Configuration parameters of Y-shaped end-flared pier

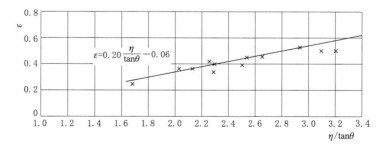

Key

×   actual project data
ε contraction ratio
η bottom position parameter of end-flared pier end section
θ lateral contraction angle

Figure B.0.2-2 $\varepsilon - \eta/\tan\theta$ relation

Appendix C Dissipator Floor Stability Calculation

C.1 Calculation Conditions and Load Combinations

C.1.1 Dissipator floor stability shall be checked for the ultimate limit state.

C.1.2 The calculation conditions and load combinations for dissipator floor stability shall be in accordance with Table C.1.2.

Table C.1.2 Calculation conditions and load combinations for dissipator floor stability

Design situation	Load combination	Calculation condition	Load					Remarks
			Self-weight	Time-averaged pressure on upper surface	Uplift pressure	Pressure fluctuation on upper surface	Foundation anchorage force	
Persistent	Fundamental combination	Design condition	√	√	√	√	√	Design flood discharge
		Frequent flood condition	√	√	√	√	√	–
Transient		Maintenance condition	√	–	√	–	√	–
Accidental	Accidental combination	Check condition 1	√	√	√	√	√	Check flood discharge
		Check condition 2	√	√	√	√	√	Drainage system fails when discharging design or frequent flood

C.2 Load Calculation Principle

C.2.1 The self-weight of dissipator floor may be calculated in accordance with the current national standard GB/T 51394, *Standard for Load on Hydraulic Structures*.

C.2.2 In the case of flood discharge, the hydrodynamic pressure acting on the dissipator floor may be divided into time-averaged pressure and fluctuation pressure in accordance with the current national standard GB/T 51394, *Standard for Load on Hydraulic Structures*. The time-averaged pressure may be estimated by hydrostatic pressure of the calculated water surface profile or the measured water surface profile in the model test. The fluctuation pressure may be calculated by the velocity head of the calculation section and the corresponding fluctuation pressure coefficient in accordance with the current national standard GB/T 51394, *Standard for Load on Hydraulic Structures*. For important projects or projects with complex hydraulic conditions, the hydrodynamic pressure on the dissipator floor shall be determined by hydraulic model tests.

C.2.3 The uplift pressure acting on the bottom surface of the floor shall be determined based on the comprehensive analysis of downstream water depth, natural groundwater table, bedrock conditions, foundation interception and drainage measures and their effects, etc., and shall be calculated in accordance with the current national standard GB/T 51394, *Standard for Load on Hydraulic Structures*.

C.2.4 Anchorage force of foundation shall be calculated by the design anchorage force of single anchor bolt and the spacing of anchor bolts. The type, material and length of anchor bolts shall be analyzed and determined by comprehensively considering the hydrodynamic characteristics, bedrock conditions, construction and related structural requirements of different parts of the energy dissipator.

C.3 Stability Check of Dissipator Floor

C.3.1 The structural coefficients for fundamental combination and accidental combination are both taken as 1.05 when checking the stability of dissipator floor.

C.3.2 The partial factor for load shall be taken in accordance with Table C.3.2.

Table C.3.2 Partial factor for load

S.N.	Load	Partial factor		
		Symbol	Value	Remarks
1	Self-weight of floor	γ_{G1}	0.95	–
2	Time-averaged pressure	γ_{Q1}	0.95	–
3	Fluctuation pressure	γ_{Q2}	1.3	–

Table C.3.2 *(continued)*

S.N.	Load		Partial factor		
			Symbol	Value	Remarks
4	Uplift pressure	Seepage pressure	γ_{Q31}	1.2	–
		Buoyancy pressure	γ_{Q32}	1.0	–
		Uplift pressure (with pumping drainage)	γ_{Q33}	1.1	No distinction for seepage pressure and buoyancy pressure
5	Anchorage force of foundation		γ_{G2}	0.95	–

C.3.3 The stability of dissipator floor shall be calculated by the following formulae:

$$\frac{1}{\gamma_d}R(\cdot) \geq \gamma_0 \varphi S(\cdot) \tag{C.3.3-1}$$

$$R(\cdot) = \gamma_{G1}G_1 + \gamma_{Q1}Q_1 + \gamma_{G2}G_2 \tag{C.3.3-2}$$

$$S(\cdot) = \gamma_{Q2}Q_2 + \gamma_{Q3}Q_3 \tag{C.3.3-3}$$

$$G_1 = \gamma_c Z A \tag{C.3.3-4}$$

$$Q_1 = p_{tr}A \tag{C.3.3-5}$$

$$Q_2 = \beta_m P_{fr} A \tag{C.3.3-6}$$

where

γ_d is the structural coefficient;

$R(\cdot)$ is the action effect function;

γ_0 is the importance factor of structure, which may be taken as 1.1, 1.0 and 0.9 for structures and components with structural safety of Level Ⅰ, Level Ⅱ and Level Ⅲ, respectively;

φ is the design situation coefficient, which may be taken as 1.00, 0.95 and 0.85 for persistent, transient and accidental situations, respectively;

$S(\cdot)$ is the resistance function;

γ_{G1} is the partial factor for self-weight of floor;

G_1 is the characteristic value of self-weight of floor (kN);

γ_{Q1} is the partial factor for time-averaged pressure;

Q_1 is the characteristic value of the time-averaged pressure on the upper surface of floor (kN);

γ_{G2} is the partial factor for foundation anchorage force;

G_2 is the effective characteristic value of foundation anchorage force (kN);

γ_{Q2} is the partial factor for fluctuation pressure;

Q_2 is the characteristic value of fluctuation pressure acting on the upper surface of floor (kN);

γ_{Q3} is the partial factor for uplift pressure;

Q_3 is the characteristic value of uplift pressure on the bottom surface of floor including seepage pressure and buoyancy pressure (kN);

γ_c is the unit weight of floor (kN/m³);

Z is the thickness of floor (m);

A is the calculation area of floor (m²);

p_{tr} is the representative value of time-averaged pressure acting on upper surface of floor (kN/m²);

β_m is the area homogenization coefficient;

P_{fr} is the representative value of fluctuation pressure acting on upper surface of floor (kN/m²).

Appendix D Hydraulic Calculation for Ski-Jump Energy Dissipation

D.1 Hydraulic Calculation of the Equal-Width Solid Flip Bucket

D.1.1 The average velocity at the top section of an equal-width solid flip bucket may be calculated as follows:

1. When the discharge chute is long, the water depth on the top of sill may be calculated by the water surface profile, and this water depth may be used to derive the average velocity at the section.

2. When an overflow weir is in front of the flip bucket, the average velocity of equal-width solid flip bucket (Figure D.1.1) can be calculated by the following formula:

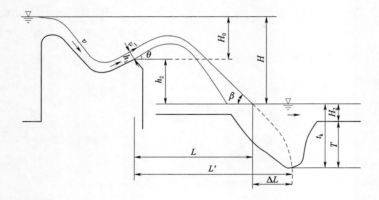

Key

- v average velocity at flip bucket top section
- v_1 water surface velocity at the top of flip bucket
- h_1 water depth at the top of flip bucket in normal direction
- θ trajectory angle of flip bucket
- h_2 elevation difference between the flip bucket top and the downstream water surface
- H_0 head at the top of flip bucket
- H difference between upstream and downstream water levels
- β angle between outer edge of water nappe and downstream water surface
- L distance between the flip bucket top and the impact point of the outer edge of water nappe
- L' distance between the flip bucket top and the lowest point of scour hole
- ΔL horizontal distance between the impact point of the outer edge of the water nappe and the lowest point of scour hole
- t_k maximum water cushion depth from the downstream water surface to the scour hole bottom
- H_2 water depth downstream of scour hole
- T depth of scour hole

Figure D.1.1 Water nappe of equal-width solid flip bucket

$$v = \varphi\sqrt{2g(H_0 - h_1\cos\theta)} \tag{D.1.1}$$

where

- v is the average velocity at flip bucket top section (m/s);
- φ is the velocity coefficient, which may be taken as 0.95 in estimate;
- H_0 is the head at the top of flip bucket (m);
- h_1 is the average water depth at the top of flip bucket in normal direction (m);
- θ is the trajectory angle of flip bucket (°).

D.1.2 The trajectory length of the outer edge of water nappe of equal-width solid flip bucket can be estimated by the following formulae:

$$L' = L + \Delta L \tag{D.1.2-1}$$

$$L = \frac{v_1^2}{g}\cos\theta\left[\sin\theta + \sqrt{\sin^2\theta + 2g(h_1\cos\theta + h_2)/v_1^2}\right] \tag{D.1.2-2}$$

$$t_k = T + H_2 \tag{D.1.2-3}$$

$$\tan\beta = \sqrt{\tan^2\theta + \frac{2g(h_2 + h_1\cos\theta)}{v_1^2\cos^2\theta}} \tag{D.1.2-4}$$

$$\Delta L = t_k \cot\beta \tag{D.1.2-5}$$

where

- L' is the horizontal distance between the flip bucket top and the lowest point of scour hole (m);
- L is the horizontal distance between the flip bucket top and the impact point of the outer edge of water nappe (m);
- ΔL is the horizontal distance between the impact point of the outer edge of water nappe and the lowest point of scour hole (m);
- v_1 is the water surface velocity at the flip bucket top (m/s), which may be taken as 1.1 times the average velocity v of the section at the flip bucket;
- g is the acceleration of gravity (m/s²);
- t_k is the maximum water cushion depth from the downstream water surface to the scour hole bottom (m);
- h_2 is the elevation difference between the flip bucket top and the downstream water surface (m);

T is the depth of scour hole (m);

θ is the trajectory angle of flip bucket (°);

β is the angle between the outer edge of water nappe and the downstream water surface (°).

D.1.3 The maximum water cushion depth of the scour hole for the water nappe of the equal-width solid flip bucket can be estimated by the following formulae:

$$t_k = Kq^{0.5}H^{0.25} \tag{D.1.3-1}$$

$$H_1 = h_1\cos\theta + h_2 + \alpha\frac{v^2}{2g} \tag{D.1.3-2}$$

where

t_k is the maximum water cushion depth from the downstream water surface to the scour hole bottom (m);

K is the erosion coefficient, which may be taken in accordance with Table D.1.3 for bedrock;

q is the unit discharge at the top of flip bucket [m³/(s · m)];

H is the difference between upstream and downstream water levels (m). If the head loss in front of the flip bucket is relatively high, H is equal to the residual energy head H_1 at the top section of flip bucket, which is calculated by Formula (D.1.3-2);

H_1 is the residual energy head at the top section of flip bucket (m);

α is the kinetic energy correction factor, which may be taken as 1.00 to 1.05.

Table D.1.3 Erosion coefficient K of bedrock

Erodibility		Very hard to erode	Hard to erode	Easy to erode	Very easy to erode
Joint/ fissure	Spacing (cm)	> 150	50 - 150	20 - 50	< 20
	Developing degree	Undeveloped, 1 to 2 joint/ fissure sets, regular	Underdeveloped, 2 to 3 joint/ fissure sets, X-shaped, relatively regular	Developed, more than 3 joint/fissure sets, X-shaped or star-shaped	Well developed, more than 3 joint/fissure sets, highly fractured, very closely jointed rock mass

Table D.1.3 *(continued)*

Erodibility		Very hard to erode	Hard to erode	Easy to erode	Very easy to erode
Bedrock structure	Integrity degree	Massive	Blocky	Closely jointed	Very closely jointed
	Structure type	Massive	Slightly folded	Moderately folded	Intensely folded
	Fissure properties	Mostly primitive or structural, mostly closed, short in persistence	Mainly structural, mostly closed, some slightly open, little filling, well cemented	Mainly structural or weathered, mostly slightly open, some open, some filled with clay, poorly cemented	Mainly weathered or structural, slightly open or open, partially some filled with clay, very poorly cemented
K	Range	0.6 - 0.9	0.9 - 1.2	1.2 - 1.6	1.6 - 2.0
	Average value	0.8	1.1	1.4	1.8

NOTE This table is applicable to the water nappe impact angle between 30° and 70°.

D.2 Hydraulic Calculation for Slit Flip Bucket

D.2.1 The trajectory length of water nappe outer edge of the slit flip bucket (Figure D.2.1) can be estimated by the following formulae:

$$L_1 = \frac{v_m^2}{g} \cos\theta_m \left[\sin\theta_m + \sqrt{\sin^2\theta_m + 2g(h_1 + h_2)/v_m^2} \right] \quad \text{(D.2.1-1)}$$

$$v_m = \varphi\sqrt{2gH} \quad \text{(D.2.1-2)}$$

$$\theta_m = \tan^{-1}\left[\frac{1}{\sqrt{1 + \frac{2g(h_1 + h_2)}{v_m^2}}} \right] \quad \text{(D.2.1-3)}$$

where

L_1 is the trajectory distance from the flip bucket top to the impact point of water nappe outer edge (m);

v_m is the velocity of the outer edge of the water nappe at the top of the slit flip bucket (m/s);

θ_m is the trajectory angle of water surface velocity v_m at the flip bucket top (°), which may be taken as 40° to 45° in estimate or may be taken as the trajectory angle of the longest trajectory length;

h_1 is the vertical water depth at the top of flip bucket (m);

h_2 is the elevation difference from the flip bucket top to the downstream water surface (m);

φ is the velocity coefficient, which may be taken as 0.80 to 0.90;

H is the difference between upstream and downstream water levels (m).

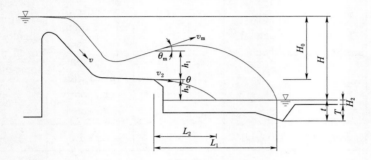

Key

v average velocity at cross section

v_m trajectory velocity of the outer edge of the water nappe at the top of slit flip bucket

θ_m trajectory angle corresponding to the water surface velocity at the top of flip bucket

h_1 vertical water depth at the top of flip bucket

v_2 trajectory velocity at the top of the flip bucket

θ trajectory angle of flip bucket

h_2 elevation difference from the flip bucket top to the downstream water surface

L_2 horizontal distance from the flip bucket top to the impact point of inner edge of water nappe

L_1 horizontal distance from the flip bucket top to the impact point of outer edge of water nappe

H_0 head at the top of flip bucket

H difference between upstream and downstream water levels

t elevation difference between the downstream water surface and the lowest point of scour hole

H_2 water depth downstream of scour hole

T depth between the lowest point of scour hole and the downstream riverbed

Figure D.2.1 Water nappe of slit flip bucket

D.2.2 The trajectory length of water nappe inner edge of the slit flip bucket can be estimated by the following formulae:

$$L_2 = \frac{v_2^2}{g} \cos\theta \left(\sin\theta + \sqrt{\sin^2\theta + 2gh_2/v_2^2} \right) \qquad (D.2.2\text{-}1)$$

$$v_2 = \varphi\sqrt{2g(H-h_2)} \qquad (D.2.2\text{-}2)$$

where

- L_2 is the trajectory length from the flip bucket top to the impact point of inner edge of water nappe (m);
- v_2 is the velocity at the top of flip bucket (m/s);
- θ is the trajectory angle of flip bucket (°);
- φ is the velocity coefficient, which may be taken as 0.65 to 0.75;
- H is the difference between upstream and downstream water levels (m).

D.2.3 The maximum depth of water cushion in scour hole can be estimated by the following formulae:

$$t = Kq^{0.5}H^{0.25}\varepsilon^n \qquad (D.2.3\text{-}1)$$

$$\varepsilon = \frac{b}{B} \qquad (D.2.3\text{-}2)$$

where

- t is the maximum water cushion depth between the downstream water surface and the lowest point of scour hole (m), which is taken as H_2 if t is smaller than H_2;
- K is the erosion coefficient, which may be taken from Table D.1.3;
- q is the unit discharge before entering the slit [m³/(s · m)], which is equivalent to the unit discharge of equal-width flip bucket;
- H is the difference between upstream and downstream water levels (m), which is taken as the residual energy head at the top section of flip bucket if the head loss in front of the flip bucket is relatively large;
- ε is the contraction ratio of slit flip bucket;
- n is the exponent, which shall be determined by tests, or may be taken as 1/3 to 1/2 in estimate, the larger value for smaller contraction ratio, or vice versa;
- b is the width of the exit section of flip bucket (m);
- B is the width of the entrance section of flip bucket (m).

D.2.4 The horizontal distance from the lowest point of scour hole to the flip bucket top can be estimated as follows:

$$L' \approx L_1 \tag{D.2.4}$$

where

L' is the horizontal distance from the lowest point of scour hole to the flip bucket top (m).

Appendix E Hydraulic Calculation for Surface Flow Energy Dissipation

E.0.1 The height of bucket lip may be calculated as follows:

1 The minimum height $a_{\min}$ of bucket lip can be calculated by the following formulae:

$$\frac{a_{\min}}{h_k} = 0.186\left(\frac{h_1}{h_k}\right)^{-1.75} \qquad (E.0.1\text{-}1)$$

$$h_k = \sqrt[3]{\frac{\alpha q^2}{g}} \qquad (E.0.1\text{-}2)$$

where

$a_{\min}$ is the minimum height of bucket lip (m);

h_k is the critical water depth corresponding to unit discharge q (m), which is calculated by Formula (E.0.1-2) for a rectangular section;

h_1 is the contraction water depth on bucket lip (m), which is calculated by the upstream total head E_0 above the bucket bottom;

α is the kinetic energy correction factor, which may be approximately 1.0;

q is the unit discharge [m³/(s · m)];

g is the acceleration of gravity (m/s²).

2 The height limit a_1 of bucket lip to form the free surface flow regime can be calculated by the following formulae:

$$a_1 = h_{\text{ocp}} - 2h_1 - h_t + 2\sqrt{h_t^2 - A} \qquad (E.0.1\text{-}3)$$

$$h_{\text{ocp}} = \frac{1}{3}\left(1 + \sqrt{6Fr_1^2 + 1}\right)h_1 \qquad (E.0.1\text{-}4)$$

$$A = 2Fr_1^2 h_1^3\left(\frac{\alpha_1}{h_1} - \frac{\alpha_t}{a + h_1}\right) \qquad (E.0.1\text{-}5)$$

$$Fr_1 = \frac{v}{\sqrt{gh_1}} \qquad (E.0.1\text{-}6)$$

where

a_1 is the height limit of bucket lip to form the free surface flow regime (m);

h_{ocp} is the critical head increment (m);

h_t is the downstream water depth (m);

Fr_1 is the Froude number on the bucket lip;

α_1, α_t are the momentum correction factors, which may be taken as 1.0;

v is the flow velocity on the bucket lip (m/s);

a is the height of bucket lip (m).

3   The height limit a_3 of bucket lip to form the submerged surface flow regime can be calculated by the following formulae:

$$a_3 = -h_{ocp} + \sqrt{(h_{ocp} - h_1)h_{ocp} + h_t^2 - A} \qquad \text{(E.0.1-7)}$$

$$A = 2Fr_1^2 h_1^3 \left(\frac{\alpha_1}{h_1} - \frac{\alpha_t}{a + h_{ocp}} \right) \qquad \text{(E.0.1-8)}$$

where

a_3 is the height limit of bucket lip to form the submerged surface flow regime (m).

E.0.2   The water depth limits can be calculated by the following formulae:

$$\frac{h_{t1}}{h_k} = 0.84 \frac{a}{h_k} - 1.48 \frac{a}{P} + 2.24 \qquad \text{(E.0.2-1)}$$

$$\frac{h_{t2}}{h_k} = 1.16 \frac{a}{h_k} - 1.81 \frac{a}{P} + 2.38 \qquad \text{(E.0.2-2)}$$

$$\frac{h_{t3}}{h_k} = \left(4.33 - 4.00 \frac{a}{P} \right) \frac{a}{h_k} + 0.9 \qquad \text{(E.0.2-3)}$$

where

h_{t1} is the limit water depth of the first critical flow regime (m);

a is the height of bucket lip (m);

P is the height of weir above downstream riverbed (m);

h_{t2} is the limit water depth of the second critical flow regime (m);

h_{t3} is the limit water depth of the third critical flow regime (m).

Translator's Annotation: The limit water depth of the first critical flow regime, h_{t1}, refers to the minimum downstream water depth where the free surface flow regime occurs; the limit water depth of the second critical flow regime, h_{t2}, refers to the minimum downstream water depth where the free surface flow or mixed surface flow transforms *to submerged surface flow; the limit water depth of the third critical flow regime, h_{t3}, refers to the maximum downstream water depth where the submerged mixed surface flow or submerged surface flow regime maintains without forming the return underflow.*

Appendix F Hydraulic Calculation for Inside-Tunnel Energy Dissipation

F.1 Diameter Estimation of Swirl Shaft and Horizontal Vortex Tunnel

In the preliminary design, the diameter of a swirl shaft or horizontal vortex tunnel can be estimated by the following formula:

$$D = k\left(\frac{Q_m^2}{g}\right)^{0.2} \tag{F.1.1}$$

where

- D is the diameter of swirl shaft or horizontal vortex tunnel (m);
- k is the coefficient, taken as 1.00 to 1.05;
- Q_m is the maximum design discharge flow (m³/s);
- g is the acceleration of gravity (m/s²).

F.2 Head Loss Estimation of Orifice Plate

F.2.1 The shape coefficient of orifice plate end may be determined by the shape coefficient curves of Type A and Type B orifice plate ends (Figure F.2.1), or may be calculated as follows:

1 If the ratio of the edge radius of orifice plate end to the hole diameter of orifice plate is not greater than 0.12, the shape coefficient of Type A orifice plate end can be calculated by the following formula:

$$\zeta' = 0.5 e^{-15r/d} \tag{F.2.1-1}$$

where

- ζ' is the shape coefficient of orifice plate end;
- r is the edge radius of orifice plate end (m);
- d is the hole diameter of orifice plate (m).

2 If the ratio of the edge radius of orifice plate end to the hole diameter of orifice plate is not greater than 0.06, the shape coefficient of Type B orifice plate end can be calculated by the following formula:

$$\zeta' = 0.5 - 8.3(r/d) + 200(r/d)^3 \tag{F.2.1-2}$$

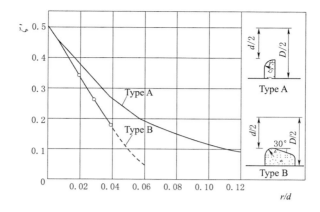

Key

o   test data
---   calculated by the fitting formula
ζ' shape coefficient of orifice plate end
d hole diameter of orifice plate
D diameter of the tunnel section between orifice plates
r edge radius of orifice plate end

Figure F.2.1   Shape coefficient curves of Type A and Type B orifice plate ends

F.2.2   The relationship between the resistance coefficient and the shape of the mth orifice plate can be estimated by the following formulae:

$$\zeta_m = \left[1 + \sqrt{\zeta'(1-\beta_m^2)} - \beta_m^2\right]^2 / \beta_m^4 \qquad \text{(F.2.2-1)}$$

$$\beta_m = \frac{d_m}{D} \qquad \text{(F.2.2-2)}$$

where

ζ_m is the resistance coefficient of orifice plate;

β_m is the ratio of hole diameter;

d_m is the hole diameter of the mth orifice plate (m);

D is the diameter of tunnel section between orifice plates (m), i.e. the diameter of tunnel section between $0.5D$ upstream and $2.5D$ downstream of the mth orifice plate.

F.2.3   The total head loss of the mth orifice plate section can be estimated by the following formulae:

$$h_{wm} = \left(Z_m + \frac{P_m}{\gamma} + \alpha_m \frac{\overline{V_m^2}}{2g}\right) - \left(Z_{m+1} + \frac{P_{m+1}}{\gamma} + \alpha_{m+1} \frac{\overline{V_{m+1}^2}}{2g}\right) \qquad \text{(F.2.3-1)}$$

$$h_{wm} = h_{jm} + h_{fm} = \zeta_m \frac{\overline{V^2}}{2g} + \lambda_m \frac{l_m \overline{V^2}}{4R2g} \tag{F.2.3-2}$$

where

- h_{wm} is the total head loss of the mth orifice plate section (m);
- Z_m is the crown elevation of the tunnel at 0.5D upstream of the mth orifice plate (m);
- P_m is the pressure on the crown of the tunnel at 0.5D upstream of the mth orifice plate (Pa);
- γ is the unit weight of water, which may be taken as 9,800 N/m³;
- α_m is the kinetic energy correction factor at 0.5D upstream of the mth orifice plate, which may be taken as 1.00 to 1.05;
- $\overline{V}_m$ is the average velocity at 0.5D upstream of the mth orifice plate (m/s);
- Z_{m+1} is the crown elevation of the tunnel 2.5D downstream of the mth orifice plate (m);
- P_{m+1} is the pressure on the crown of the tunnel at 2.5D downstream of the mth orifice plate (Pa);
- α_{m+1} is the kinetic energy correction factor at 2.5D downstream of the mth orifice plate, which may be taken as 1.00 to 1.05;
- $\overline{V}_{m+1}$ is the average velocity at 2.5D downstream of the mth orifice plate (m/s);
- h_{jm} is the local head loss of orifice plate (m);
- h_{fm} is the frictional head loss of orifice plate section (m);
- ζ_m is the resistance coefficient of orifice plate;
- λ_m is the frictional loss factor;
- l_m is the length of the mth orifice plate section (m);
- R is the hydraulic radius of orifice plate section (m);
- $\overline{V}$ is the average velocity in orifice plate section (m/s).

F.3 Head Loss Estimation of Thick Orifice

F.3.1 The head loss estimation of thick orifice (Figure F.3.1) shall cover the head loss due to sudden expansion and contraction, and the estimated results shall be corrected based on the test data.

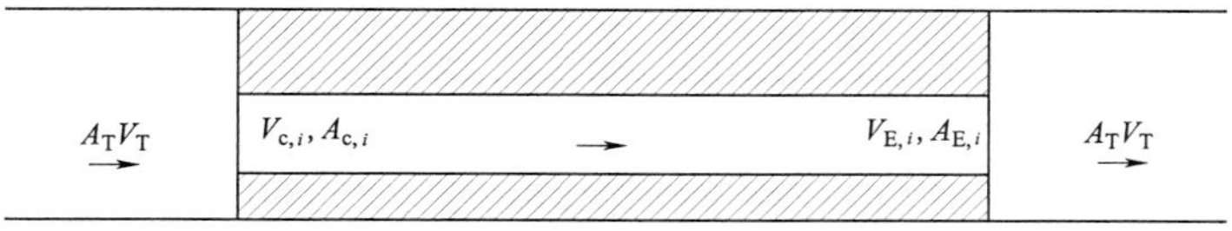

Key

A_T sectional area of spillway tunnel

V_T average velocity in spillway tunnel

$V_{c,i}$ average velocity at the inlet of the ith thick orifice

$A_{c,i}$ sectional area at the inlet of the ith thick orifice

$V_{E,i}$ average velocity at the outlet of the ith thick orifice

$A_{E,i}$ sectional area at the outlet of the ith thick orifice

Figure F.3.1 Thick orifice

F.3.2 The sudden-contraction head loss of the ith thick orifice can be estimated as follows:

$$\Delta H_{c,i} = \left[\frac{A_T}{A_{c,i}} - \left(\frac{A_{c,i}}{A_T}\right)^{0.25}\right]^2 \frac{V_T^2}{2g} \tag{F.3.2}$$

where

$\Delta H_{c,i}$ is the sudden-contraction head loss of the ith thick orifice (m);

A_T is the sectional area of spillway tunnel (m^2);

$A_{c,i}$ is the sectional area at the inlet of the ith thick orifice (m^2);

V_T is the average velocity in spillway tunnel (m/s).

F.3.3 The sudden-expansion head loss of the ith thick orifice can be estimated as follows:

$$\Delta H_{E,i} = 0.75\left[\frac{A_T}{A_{E,i}} - \left(\frac{A_{E,i}}{A_T}\right)^{0.25}\right]^2 \frac{V_T^2}{2g} \tag{F.3.3}$$

where

$\Delta H_{E,i}$ is the sudden-expansion head loss of the ith thick orifice (m);

$A_{E,i}$ is the sectional area at the outlet of the ith thick orifice (m^2).

Appendix G Scour Calculation of Downstream River Channel

G.0.1 The anti-scouring design of riverbed should meet the following requirements:

1 If $\sqrt{q\sqrt{\Delta H}} = 1 \sim 9$, the apron extension length can be calculated by the following formula:

$$L_p = k_s \sqrt{q\sqrt{\Delta H}} \tag{G.0.1-1}$$

where

L_p is the apron extension length (m);

k_s is the coefficient for calculating apron extension length, which may be taken in accordance with Table G.0.1;

q is the unit discharge at the end of stilling basin [m³/(s · m)];

ΔH is the difference between upstream and downstream water levels (m).

Table G.0.1 Coefficient for calculating apron extension length

Riverbed soil	Hard clay	Silty clay	Medium sand, coarse sand, silty loam	Silty sand, fine sand
k_s	7 - 8	9 - 10	11 - 12	13 - 14

2 The scour depth at the end of the apron extension can be calculated by the following formula:

$$d = 1.1 \frac{q}{v_0} - h_s \tag{G.0.1-2}$$

where

d is the scour depth of the riverbed at the end of the apron extension (m);

q is the unit discharge at the end of the apron extension [m³/(s · m)];

v_0 is the allowable non-scouring velocity of riverbed soil (m/s);

h_s is the water depth at the end of the apron extension (m).

G.0.2 The scour depth of riverbank slope can be calculated by the following formulae:

$$h_s = H_0 \left[\left(\frac{U_{cp}}{U_c} \right)^n - 1 \right] \qquad (G.0.2\text{-}1)$$

$$U_{cp} = U \frac{2\eta}{1+\eta} \qquad (G.0.2\text{-}2)$$

where

- h_s is the local scour depth (m);
- H_0 is the water depth at the scouring position (m);
- U_{cp} is the average velocity of flow normal to bank (m/s);
- U_c is the sediment-moving incipient velocity (m/s);
- n is the coefficient relating to the shape of the protective riverbank slope on the plane, which may be taken as 1/6 to 1/4;
- U is the approach velocity (m/s);
- η is the uniformity coefficient for flow velocity, which may be taken in accordance with Table G.0.2.

Table G.0.2 Uniformity coefficient for flow velocity

Intersection angle between flow direction and bank slope α (°)	≤ 15	20	30	40	50	60	70	80	90
η	1.00	1.25	1.50	1.75	2.00	2.25	2.50	2.75	3.00

G.0.3 The sediment-moving incipient velocity may be calculated based on the particle properties as follows:

1. The incipient velocity of cohesive soil and sand can be calculated by the following formula:

$$U_c = \left(\frac{H_0}{d_{50}} \right)^{0.14} \sqrt{17.6 \frac{\gamma_s - \gamma}{\gamma} d_{50} + 0.0000000605 \frac{10 + H_0}{d_{50}^{0.72}}} \qquad (G.0.3\text{-}1)$$

where

- U_c is the incipient velocity of cohesive soil or sand (m/s);
- d_{50} is the median grain size of cohesive soil or sand (m);
- γ_s is the unit weight of sand (kN/m³);
- γ is the unit weight of water (kN/m³).

2. The incipient velocity of cobbles can be calculated by the following

formula:

$$U_c = 1.08\sqrt{gd_{50}\frac{\gamma_s - \gamma}{\gamma}}\left(\frac{H_0}{d_{50}}\right)^{\frac{1}{7}} \tag{G.0.3-2}$$

where

U_c is the incipient velocity of cobbles (m/s);

d_{50} is the median grain size of cobbles (m).

Appendix H Empirical Allowable Non-scouring Velocity for Downstream Channel

H.0.1 The empirical allowable non-scouring velocity in rock river channel may be taken from Table H.0.1.

Table H.0.1 Empirical allowable non-scouring velocity in rock river channel

Rock	Water depth (m)			
	0.4	1.0	2.0	3.0
	Allowable non-scouring velocity (m/s)			
Conglomerate, marl, shale	2.0	2.5	3.0	3.5
Porous limestone, dense conglomerate, stratified limestone, calcareous sandstone, dolomitic limestone	3.0	3.5	4.0	4.5
Dense limestone, siliceous limestone, marble	4.0	5.0	5.5	6.0
Granite, diabase, basalt, quartz porphyry	15.0	18.0	20.0	22.0

H.0.2 The allowable non-scouring velocity in cohesive soil river channel may be taken from Table H.0.2.

Table H.0.2 Allowable non-scouring velocity in cohesive soil river channel

Soil	Value of allowable non-scouring velocity (m/s)
Light loam	0.60 - 0.80
Medium loam	0.65 - 0.85
Heavy loam	0.70 - 0.95
Clay	0.75 - 1.00

NOTE This table is applicable to the hydraulic radius of 1.0 m. When the hydraulic radius is not 1.0 m, the values in the table shall be adjusted by multiplying the hydraulic radius to the power α, which may be 1/4 to 1/3 for the loose loam or clay river channel, or 1/5 to 1/4 for the medium dense to dense loam or clay river channel.

H.0.3 The allowable non-scouring velocity in non-cohesive soil river channel may be taken from Table H.0.3.

Table H.0.3 Allowable non-scouring velocity in non-cohesive soil river channel

Soil	Grain size (mm)	Water depth (m)			
		0.4	1.0	2.0	≥ 3.0
		Allowable non-scouring velocity (m/s)			
Silt	0.005 - 0.050	0.12 - 0.17	0.15 - 0.21	0.17 - 0.24	0.19 - 0.26
Fine sand	0.050 - 0.250	0.17 - 0.27	0.21 - 0.32	0.24 - 0.37	0.26 - 0.40
Medium sand	0.250 - 1.000	0.27 - 0.47	0.32 - 0.57	0.37 - 0.65	0.40 - 0.70
Coarse sand	1.000 - 2.500	0.47 - 0.53	0.57 - 0.65	0.65 - 0.75	0.70 - 0.80
Fine gravel	2.500 - 5.000	0.53 - 0.65	0.65 - 0.80	0.75 - 0.90	0.80 - 0.95
Medium gravel	5.000 - 10.00	0.65 - 0.80	0.80 - 1.00	0.90 - 1.10	0.95 - 1.20
Large gravel	10.00 - 15.00	0.80 - 0.95	1.00 - 1.20	1.10 - 1.30	1.20 - 1.40
Fine cobble	15.00 - 25.00	0.95 - 1.20	1.20 - 1.40	1.30 - 1.60	1.40 - 1.80
Medium cobble	25.00 - 40.00	1.20 - 1.50	1.40 - 1.80	1.60 - 2.10	1.80 - 2.20
Large cobble	40.00 - 75.00	1.50 - 2.00	1.80 - 2.40	2.10 - 2.80	2.20 - 3.00
Small boulder	75.00 - 100.0	2.00 - 2.30	2.40 - 2.80	2.80 - 3.20	3.00 - 3.40
Medium boulder	100.0 - 150.0	2.30 - 2.80	2.80 - 3.40	3.20 - 3.90	3.40 - 4.20
Large boulder	150.0 - 200.0	2.80 - 3.20	3.40 - 3.90	3.90 - 4.50	4.20 - 4.90
Very large boulder	> 200.0	> 3.20	> 3.90	> 4.50	> 4.90

NOTE This table is applicable to the hydraulic radius of 1.0 m. When the hydraulic radius is not 1.0 m, the values in the table shall be adjusted by multiplying the hydraulic radius to the power α, which may be taken as 1/5 to 1/3.

Explanation of Wording in This Guide

1. Words used for different degrees of strictness are explained as follows in order to mark the differences in executing the requirements in this guide.

    1) Words denoting a very strict or mandatory requirement:

        "Must" is used for affirmation; "must not" for negation.

    2) Words denoting a strict requirement under normal conditions:

        "Shall" is used for affirmation; "shall not" for negation.

    3) Words denoting a permission of a slight choice or an indication of the most suitable choice when conditions permit:

        "Should" is used for affirmation; "should not" for negation.

    4) "May" is used to express the option available, sometimes with the conditional permit.

2. "Shall meet the requirements of..." or "shall comply with..." is used in this guide to indicate that it is necessary to comply with the requirements stipulated in other relative standards and codes.

List of Quoted Standards

GB 50086,	Technical Code for Engineering of Ground Anchoring and Shotcrete Support
GB 50286,	Code for Design of Levee Project
GB/T 51394,	Standard for Load on Hydraulic Structures
NB/T 35026,	Design Code for Concrete Gravity Dams
NB/T 10391,	Code for Design of Hydraulic Tunnel
DL/T 5057,	Design Specification for Hydraulic Concrete Structures
DL/T 5166,	Design Specification for River-Bank Spillway
DL 5180,	Classification & Design Safety Standard of Hydropower Projects
DL/T 5244,	Code for Normal Hydraulics Model Investigation for Hydropower & Water Resources
DL/T 5245,	Code for Hydraulic Investigation on Aeration-Cavitation Resistance for Hydropower & Water Resources
DL/T 5353,	Design Specification for Slope of Hydropower and Water Conservancy Project
DL/T 5359,	Code for Model Test of Cavitation for Hydropower & Hydraulic Engineering
SL 158,	Code for Model Test on Flow Pressure Fluctuation and Flow Induced Vibration of Hydraulic Structures
SL 265,	Design Specifications for Sluices